Jane

SATURN

NICOLE MORTILLARO

Astronomy compels the soul to look upwards
and leads us from this world to another. — PLATO

FIREFLY BOOKS

A FIREFLY BOOK

Published by Firefly Books Ltd. 2010

First printing

PUBLISHER CATALOGING-IN-PUBLICATION DATA (U.S.)
Mortillaro, Nicole.
Saturn : exploring the mystery of the ringed planet / Nicole Mortillaro.
[96] p. : photos. (chiefly col.) ; cm.
Includes index.
Summary: Featuring photos from NASA resources, Saturn examines the planet and its place in our universe with a special emphasis on the most recent discoveries of the Cassini probe.
ISBN-13: 978-1-55407-649-9 ISBN-10: 1-55407-649-8
1. Saturn (Planet) – Pictorial works. I. Title.
523.46 dc22 QB671.M678 2010

LIBRARY AND ARCHIVES CANADA CATALOGUING IN PUBLICATION
Mortillaro, Nicole, 1972-
Saturn : exploring the mystery of the ringed planet / Nicole Mortillaro.
Includes index.
ISBN 978-1-55407-649-9 ISBN-10: 1-55407-649-8
1. Saturn (Planet). 2. Saturn (Planet)--Pictorial works.
3. Cassini (Spacecraft). I. Title.
QB671.M67 2010 523.46 C2010-903263-2

Published in the United States by
Firefly Books (U.S.) Inc.
P.O. Box 1338, Ellicott Station
Buffalo, New York 14205

Published in Canada by
Firefly Books Ltd.
66 Leek Crescent
Richmond Hill, Ontario L4B 1H1

Interior design: opushouse.ca
Formatting: opushouse.ca

Printed in China

The publisher gratefully acknowledges the financial support for our publishing program by the Government of Canada through the Canada Book Fund as administered by the Department of Canadian Heritage.

CONTENTS

SATURN: AN INTRODUCTION

Ever since we first gazed up into the heavens, we have been curious about the origins of the universe around us. Before light pollution and cars and buildings, the night sky was afire with light. Humankind sought to peer deeper and ever farther into the stars that set the night sky alight, but how to travel to the farthest outreaches of space?

Most people are familiar with the term the "Big Bang," first coined by Monsignor Georges Henri Joseph Édouard Lemaître and George Gamow in the 1950s in reference to the moment when our universe took shape. It's believed that the origins of life began with a single moment 13.7 billion years ago, when elements were formed, leading to stars and planets and, of course, us. The age of our own solar system is estimated to be around 4.5 billion years old; humans have only been on Earth a few hundred thousand years. So, putting that into perspective, we are mere infants, universally speaking.

The naming of the planets comes from the Greek word *planetes*, which means "wanderers." The name makes sense, as the planets move across the sky very differently than the fixed stars. As Saturn is one of the five planets that people can see with the naked eye, our ancestors have known about its existence for thousands of years.

The Greeks had named Saturn after Kronos — a god who had eaten his children for fear of them usurping him from the throne. Later, the Romans named the planet after their god of agriculture. Then in 1610, hundreds of years later, it was first viewed through a telescope by Galileo Galilei. Galileo's telescope, though marvellous, was too crude to make out much detail. But he did see the bulges, or "ears," on either side of the planet and attributed them to two moons on either side of Saturn. As the years progressed, the astronomer noted changes it its appearance, at one point believing that Kronos had "devoured his own children." What he was actually seeing was the change in Saturn's tilt relative to Earth. In 1655, Christiaan Huygens was the first to realize that Saturn had a ring system. Four years later, he also discovered Saturn's largest moon, Titan. As time progressed, more and more information was collected on the planet, most of it done from afar, as humans were Earth-bound. But this was just the beginning.

As children are want to do, we explored and discovered, seeking to further our knowledge of our surroundings. It's taken us millions of years of evolution to be able to build that knowledge, but a true understanding of our universe didn't come into being until relatively recently. Technological advances allowed us to peer deeper than anyone could ever have imagined. Spectroscopy, x-rays and ultraviolet photography helped us to analyze the makeup of stars and planets. But still, we were relegated to far-off observation.

And then the age of rocketry came upon us. We sent men to circle our pale blue orb. Then we went even farther — we sent men to walk upon a body other than Earth. However, the moon was as far as we'd be able to go for some time, and we were all too aware of that.

But machines — complicated, fascinating machines — *could* travel the vast distances that separated the bodies in our solar system. If we could somehow use the technology available to us to help sate our curiosity and provide us with at least some of the answers we sought, perhaps we could enhance our understanding of our universe, and thus, of our own origins.

And this is what humankind has done. Some may call our voyages to the worlds around us our greatest achievement. But there is much more to come.

SATURN QUICK FACTS

Saturn, the sixth planet from the sun, is a mighty gas giant, second only to Jupiter. The planet, like its big brother Jupiter, is made up of mainly helium and hydrogen. It is so light that if a body of water big enough to hold it were discovered, the planet would bob gently on its watery surface. It is roughly 889 million miles (1.43 billion km) from Earth, yet its gaseous surface is highly reflective, making it easy to spot with the naked eye from our home planet.

Diameter (at the equator): 74,900 miles (120,540 km)
Average distance from Sun: 886,500,000 miles (1,426,725,400 km)
Length of year: 29.4 Earth years
Length of day: 10 hours and 47 minutes (approximately)

THE MOONS OF SATURN *(and counting)*

1. Mimas
2. Enceladus
3. Tethys
4. Dione
5. Rhea
6. Titan
7. Hyperion
8. Iapetus
9. Erriapus
10. Phoebe
11. Janus
12. Epimetheus
13. Helene
14. Telesto
15. Calypso
16. Kiviuq
17. Atlas
18. Prometheus
19. Pandora
20. Pan
21. Ymir
22. Paaliaq
23. Tarvos
24. Ijiraq
25. Suttungr
26. Mundilfari
27. Albiorix
28. Skathi
29. Siarnaq
30. Thrymr
31. Narvi
32. Methone
33. Pallene
34. Polydeuces
35. Daphnis
36. Aegir
37. Bebhionn
38. Bergelmir
39. Bestla
40. Farbauti
41. Fenrir
42. Fornjot
43. Hati
44. Hyrrokkin
45. Kari
46. Loge
47. Skoll
48. Surtur
49. S/2004 S7
50. S/2004 S12
51. S/2004 S13
52. S/2004 S17
53. S/2006 S1
54. S/2006 S3
55. Greip
56. Jarnsaxa
57. Tarqeq
58. S/2007 S2
59. S/2007 S3
60. Anthe
61. Aegaeon
62. S/2009 S1

Pioneers to the Heavens: Pioneer and Voyager 1 and 2

Exploring the inner planets wasn't anything new to scientists; the Venera and Mariner missions had voyaged to Mercury and Venus, and Viking had traveled to Mars. Luna and Apollo had visited the moon, with the Americans being the first to actually set foot on another body of our solar system. There had been many lessons — one hesitates to call them "mistakes" — learned along the way, as with every explosion or missed approach or lost communication adjustments were made, and our ability to get it right improved. With their successes, scientists aimed to venture even farther, to the outer planets.

The first to actually travel past the asteroid belt — a dangerous mission that had scientists holding their breath hoping that their expensive and delicate machine could maneuver through the rocks and debris without being destroyed — was Galileo, on a destination to Jupiter. And then came Pioneer 10, which also visited Jupiter. In 1973, Pioneer 11 was launched, and it too explored Jupiter, but also sent home the first close-up images of our lovely ringed planet, Saturn. With those missions proving definitively that humankind could send ambassadors of Earth to the outer solar system, the Voyager 1 and Voyager 2 satellites were devised to voyage to even farther outreaches of space. Most importantly, they traveled close to Saturn and its satellites, with the purpose of sending humanity a close-up examination of a planet that had fascinated us for millennia.

This grainy image was breathtaking to scientists back on Earth. Pioneer 11 imaged this portrait of Saturn with its mighty moon, Titan (upper left), in 1979, at a distance of 1,545,000 miles (2,486,000 km) from the planet. Pioneer 11 had been launched on April 5, 1973, with its predecessor, Pioneer 10, having been launched a year earlier. Pioneer 10's primary mission — which was successful — was to have a close-up encounter with Jupiter. Pioneer 11 arrived at its destination, Saturn, on September 1, 1979, and flew to within roughly 13,050 miles (21,000 km) of the planet, taking the first close-up pictures of this mysterious and wondrous jewel of the solar system. The spacecraft discovered two new moons and uncovered a new ring. The Pioneer satellites helped usher in a new age of deeper exploration of the farthest reaches of our solar system.

Voyager 1 lifts off in a ball of fire at Cape Canaveral, Florida, on August 20, 1977. Only a few weeks earlier, on August 20, Voyager 2 had lifted off, heading for the same destination — Saturn. The Voyager 1 and Voyager 2 satellites were designed to take advantage of an alignment of four of the outer planets that would allow them to pass closely and take unprecedented images. But costs made it too expensive to study them all, so Voyager 1 would pass close to Jupiter, while Voyager 2 would pass close to Saturn and its giant moon, Titan. Voyager 2 was launched first, with Voyager 1 launched September 5. Voyager 2 would reach Saturn on November 20, 1980.

Scientists had worked diligently for years in developing something that would be able to withstand the travels through space to provide them with a better understanding of Saturn. Pictured above is the prototype to the Voyager satellites that were sent to the ringed planet. This prototype sits atop a test platform designed to simulate the vibrations of space, located in Pasadena, California. It was important that humanity's first venture into some of the coldest and poorly understood reaches of space was successful, as it was hoped that it would lead to further exploration of our solar system.

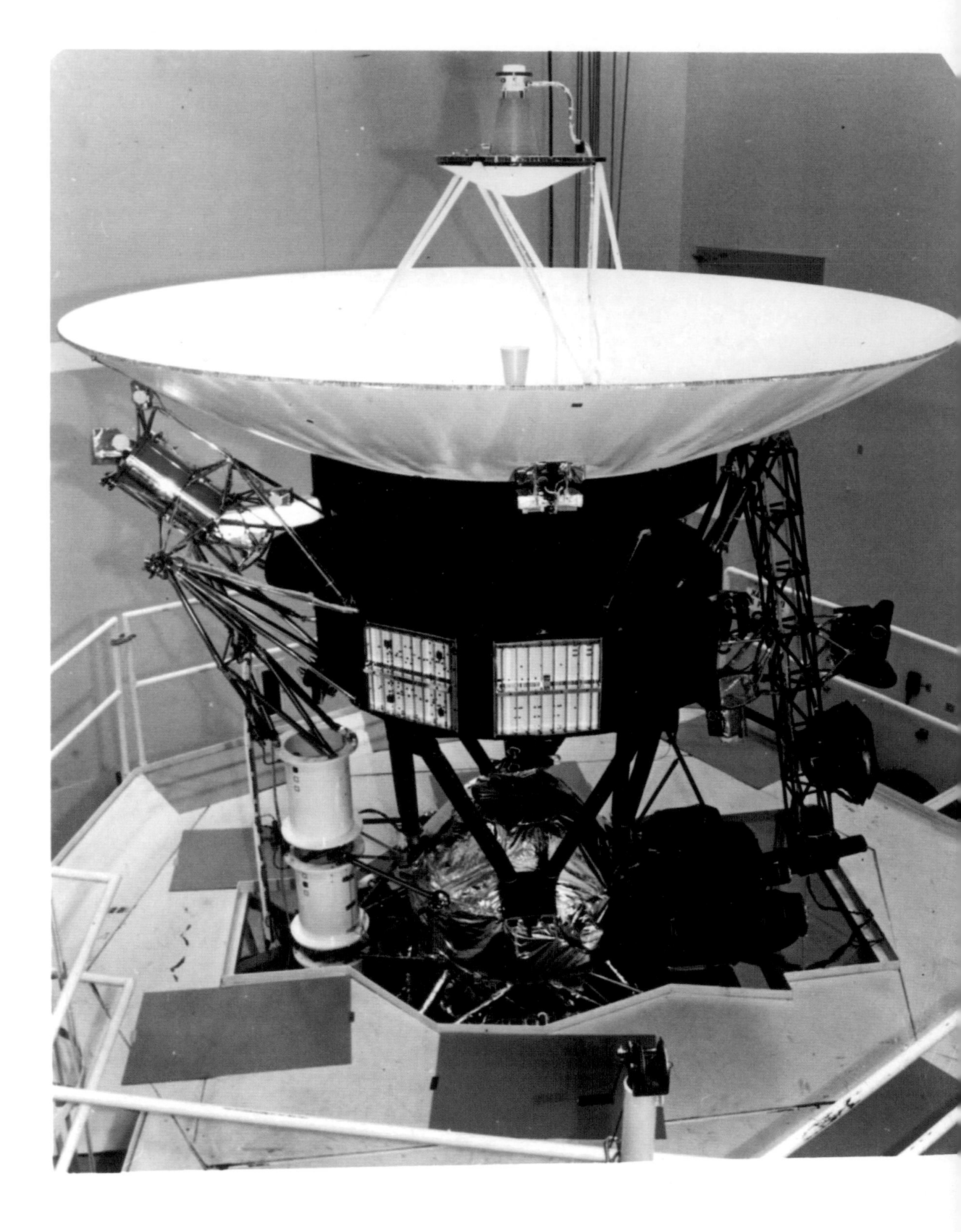

Aboard Voyager 1 and 2 are two identical records that preserve the story of Earth with the hope of carrying on the legacy of humanity into deep space. The records are 12-inch, gold-plated copper discs, and have recorded greetings in 55 different languages, as well as samples of 27 musical pieces from various cultures and eras, and 35 natural and man-made sounds. The records also have electronic information about our civilization that researchers believe advanced beings could convert into photographs and diagrams. How would these beings be able to play such a record? Each disc contains a cover with a gold-plated aluminum jacket (which provides protection against any micrometeorites that might hit the craft) and included are a cartridge and a needle.

The cover of the record (below), provides instructions on how to retrieve the information contained within it.

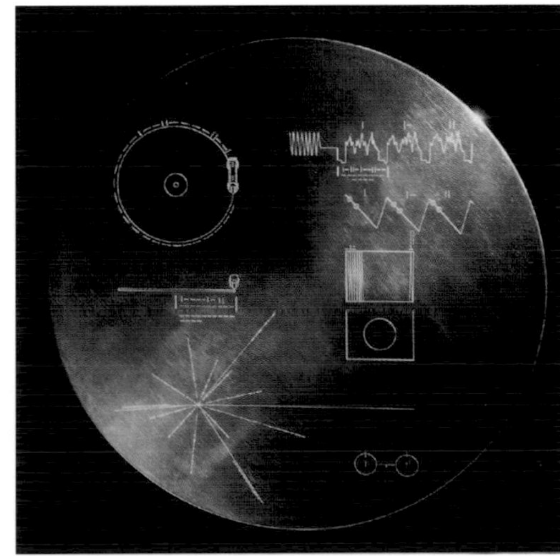

This image might not appear spectacular in any particular way, however it encapsulates the moment humankind escaped the bonds of our home and began our adventure into the deep unknown. This is the first photograph ever taken of Earth and our moon by a man-made satellite. Voyager 1 recorded this image on September 18, 1977, almost two weeks after it had set off for Saturn. It was a mere 7.25 million miles (11.66 million km) away at this point — just a relative few feet from home. The parts of Earth in sunlight are eastern Asia, the western Pacific Ocean and part of the Arctic.

This image (left) of Saturn's rings was taken by Voyager 1 at a distance of almost 5 million miles (8 million km). This is actually a composite of two images and shows almost all of the individual rings (known at the time). Before this Voyager mission, scientists had believed that the planet's rings were created by complicated gravitational interaction between the ring particles and Saturn's moons; however, this image helped them come to the conclusion that the process was too complex for that theory, begging further study.

Although it was believed that the rings of Saturn were made up of ice and dust, the interaction of these elements was poorly understood. To some degree, it still is. This image (right) was returned by Voyager 2 as it closely examined the B ring. The picture sent home presented a curious anomaly to researchers: what were these shadowy features that were present across Saturn's B ring? The features became known as "spokes" and have continued to intrigue scientists. One theory that has arisen from the analysis is that the spokes are caused by electromagnetic forces.

Saturn hangs in space with three of its moons — Tethys, Dione and Rhea — strung like a tail in front of it. This true-color image was snapped by Voyager 2 from a distance of 13 million miles (21 million km) as it approached the mighty giant. Tethys's shadow can be seen just below the rings as it makes its way across Saturn. This photograph provided scientists with one of the best images of Saturn and its pale hues, helping them see the subtleties in the planet's weather system.

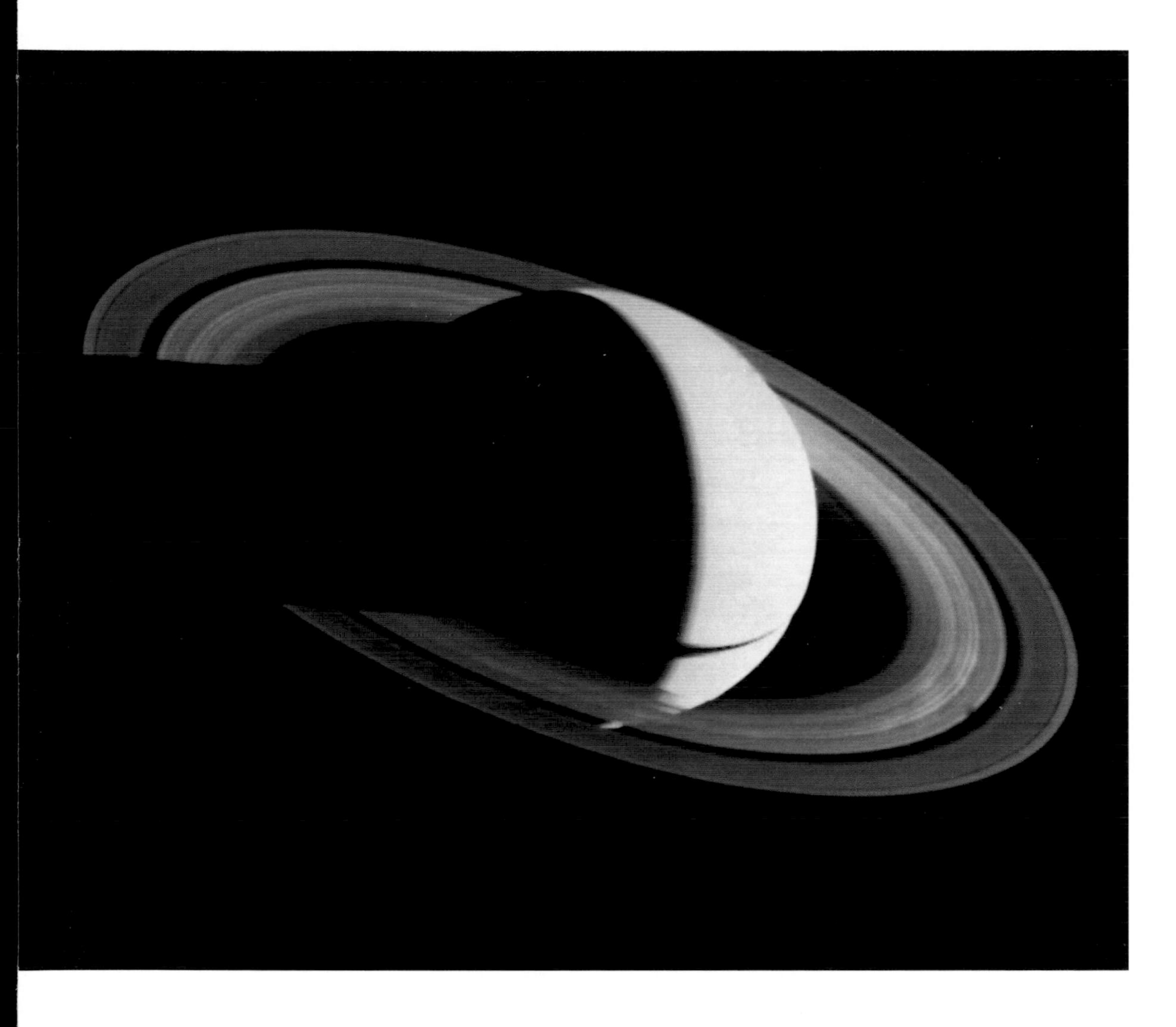

The Voyager missions had proven to scientists that Saturn possessed many secrets and that humankind still had much to learn. Here, Voyager 1 looked back at Saturn four days after it flew past the planet, at a distance of 3.3 million miles (5.3 million km). The spokes that had been discovered by the spacecraft are visible even at this distance, and Saturn's shadow falls gently across the rings as sunlight gives the planet a crescent-like appearance. This image, and many more that the Voyager spacecraft returned to Earth, convinced researchers that more missions were necessary. But it would take almost 30 years for a research mission to be sent back to the planet.

Hubble

The Hubble Space Telescope was launched in 1990 with the goal of peering ever deeper into space. The multi-million-dollar telescope was named after astronomer Edwin Hubble, who made one of the most important discoveries in astronomy — the existence of galaxies.

Although the telescope experienced some visual problems early on in its mission, Hubble was fitted with corrective lenses in an exhausting Earth-orbit repair in 1993. Even after its initial repair, the telescope has received other upgrades to help make it even more productive, and the images that it has returned from deep space have been breathtaking. The knowledge that it has provided researchers has been extensive, and its analysis continues to this day.

But Hubble was also used for what could be called backyard astronomy. It has taken beautiful portraits of Saturn and helped astronomers study the planet even further, in sharper detail than Earth-bound telescopes. Although the images it has returned are not as detailed as what would be available were it billions of kilometers closer, they still provide researchers with the unique opportunity to conduct in-depth analysis of the planet. Hubble has also captured the planet as it conducts its dance around Earth (right).

Saturn's Rings Viewed from Earth

As Saturn travels around its orbit, *Hubble* sees a different view of the tilted rings from a position near Earth. The rings nearly disappear twice during Saturn's approximately 30-year orbit, because we see them edge on and they are extremely thin relative to their diameter.

While Cassini (successor to the Voyager craft) was only a few footsteps from home on its voyage across the vast distance to Saturn, Hubble turned its new eyes — the Near Infrared Camera and Multi-Object Spectrometer (NICMOS) — toward Cassini's final destination. This image was taken on January 4, 1998, displaying the planet's reflected infrared light. The various colors tell researchers a lot about the planet: blue indicates differences in particle variations, believed to be ammonia ice crystals; green and yellow indicate haze, green being a thin layer, yellow being a thicker layer; red and orange are cloud layers, with the red indicating higher clouds; and the white coloring, near the planet's equator, indicates the densest regions, and represents storms. It is interesting to note that Earth experiences storms with the highest clouds at its equator as well. Two moons can also been seen in this image, Dione on the lower left and Tethys on the upper right. The rings have been shown in visible light.

If you've ever seen the Northern or Southern Lights here at home, you know how breathtaking they can be, as ribbons of colored light, varying from greens to reds, wind their way through the sky above. Saturn also experiences its own aurora at its poles. This series of photographs was taken on January 24, 26 and 28, 2004, as Saturn's south pole was tilted toward Earth. The aurora danced over days, and as illustrated here, varied from day to day, much as our Earthly auroras do. Scientists were intrigued to witness that Saturn's auroral display, unlike Earth's, always appears bright and lasts for a lengthy period. Scientists believe that the giant's auroral storms are caused by a stream of charged particles from the Sun, rather than by its magnetic field. Auroras here on Earth occur when electrically charged particles from the Sun interact with atoms and molecules in our upper atmosphere. Because they are electrically charged, they are attracted to the magnetic poles, making them visible most often to locations in the higher latitudes.

Hubble was able to snap these images of Saturn's enormous and enigmatic moon, Titan, as it transited (crossed) the planet in February 2009. Hubble took the images while Saturn's rings were tilted edge-on to Earth, an event that only occurs once approximately every 15 years. The first frame shows Titan (and its dark shadow) near the top of the planet. Along with Titan are Enceladus (far left, above the rings) and Dione, which appears as a tiny speck to the left of Enceladus. If you have really good eyes, you can see Mimas and its faint shadow just above the ring plane on the left. Mimas — a relatively small Saturnian moon about 246 miles (396 km) in diameter — is slightly more visible in the second frame, where it can be seen as a bright pinpoint almost behind Saturn, above the ring plane to the left. The three images capture the moons as they cross Saturn, while the planet was approximately 7.8 million miles (1.25 billion km) from Earth.

Astronomy is a hobby that requires not only superb mathematical skills, but a lot of patience. Unfortunately, the universe waits for no one and carries on at its own pace. So, when you want to examine something a little more closely, you just have to wait. Luckily, the universe does adhere to the laws of physics and math, so when astronomers wanted to understand a little more about Saturn's atmosphere, they patiently waited until the planet reached a tilt of 27 degrees away from Earth. This position put its south pole in perfect view for Hubble to snap a series of images at various wavelengths, allowing astronomers to study the giant at its maximum tilt. Thirty filters were used to snap these images on March 7, 2003. Why the various filters? The different particles in the planet's atmosphere reflect different wavelengths of light in very minute ways, causing some bands of gas to be brighter or darker. The filters also span various wavelengths, giving scientists the opportunity to determine phenomena such as cloud formation.

As Saturn journeys along its 29-year voyage around the sun, it carries out a majestic dance, nodding back and forth toward us here on Earth. These images were captured by the Hubble Space Telescope between 1996 and 2000. Just as Earth tilts on a 23-degree axis, Saturn also tilts, but at a slightly higher angle of 27 degrees, causing the giant planet to have seasons just as Earth does. The first image (lower left) was taken just after its autumnal equinox in its northern hemisphere in 1996. The final image (upper right) shows the planet tilting almost to its extreme, putting the planet in its winter solstice. These images allowed scientists to further study the rings as they tilt relative to the planet's seasons, thereby helping them to determine the composition of the rings.

Cassini-Huygens: On Approach

Following the successful missions of both the Voyager 1 and 2 spacecraft, researchers wanted to understand more about the graceful ringed jewel of the solar system. Early plans for sending another satellite to the system began in the early 1980s between NASA and several European space agencies. Then, finally, on October 15, 1997, Cassini-Huygens was Saturn-bound.

Some of the mission objectives of the spacecraft and its probe were:

1. Determining what source of heat lies within Saturn; it produces 87 percent more energy than what it actually absorbs from the sun
2. Delving deeper into understanding the origin of Saturn's rings and the reason for the rings' pale hues
3. Discovering more moons
4. Understanding why Saturn's moon Enceladus has such a smooth surface compared to other cratered satellites
5. Understanding enigmatic Titan and its atmosphere, and uncovering what its source of methane is

This photograph shows workers removing the protective cover of the Cassini craft with the attached Huygens probe (center). After years of hard work, the craft was nearing its final days on Earth. It was ready to travel the 889 million miles (1.43 billion km) through the silence of space to the second-largest planet in our solar system. The images that it would return would change forever the way we see the majestic Saturn.

Cassini and Huygens are tucked away in the mighty Titan IVB/Centaur rocket, at Launch Complex 40 at Cape Canaveral Air Station, marking the final stages before its journey toward Saturn. This image shows the Mobile Service Tower rolling away from the rocket, which marked a major milestone for the mission. Workers were anxious, as the retraction of the unit had been delayed due to some problems with ground support equipment. But the launch would be carried out, even if it was delayed slightly by two days.

After many years of planning, Cassini-Huygens soars into the early morning sky at Cape Canaveral atop the Titan IVB/Centaur rocket. After some delays, the mission began at 4:43 a.m. EDT on October 15, 1997. It would take the spacecraft seven years to traverse the deep expanse of space to reach Saturn.

When Cassini left Earth in 1997, it faced a long voyage to its final destination, and scientists weren't about to pass up the opportunity to snap photos of other planets along the way. Here, Cassini returned one of the most stunning close-up pictures ever taken of the ruler of the solar system, Jupiter. Like Saturn, Jupiter is a gas giant, with whirlwind storms whipping across its atmosphere. On December 29, 2000, Cassini captured this image at almost 6.2 million miles (10 million km) from the planet. At this distance, one can fully understand just how enormous the planet is. After gathering this data, Cassini scientists anxiously awaited the spacecraft's approach to Saturn. One could almost consider this image to be a warm-up for Cassini's ultimate destination.

Cassini took this picture of Saturn at some 69 million miles (111.4 million km) away and was only about 9 months from arriving at its final destination. Even from this distance, the distinctive bands of color in Saturn's atmosphere are visible. Like the bands of Jupiter, which are more prominent, Saturn's bands are caused by very small colored particles that interact with the white clouds of ammonia contained within its atmosphere.

Cassini's long voyage — some 600 million miles (1 billion km) from Earth — was nearing an end. It was then, on September 10, 2007, after it had its first close encounter with Saturn's moon Iapetus, that it turned its eyes toward its final destination, taking in this breathtaking view of Saturn and its satellites. This natural-color image lets us see what it might be like to view Saturn from Iapetus. This moon is the only satellite to have a significant inclination to Saturn's orbit and is the only moon that has such a view. In fact, all the other moons would see the planet's rings nearly edge-on, except Iapetus, which would see the rings on a tilt, as illustrated in this image. The moons in this image are: Dione (center, far left); Enceladus (near the left edge of the ring); Mimas (a small speck against the ring shadows at left); Rhea (upper left, in Saturn's northern hemisphere); Tethys (on the right side of the rings) and Titan (lower right).

Saturn's Grandeur

Saturn's beauty cannot be argued. The planet has existed in our mythology and minds for centuries. Compared to all the spacecraft that had come before it — Galileo, Pioneer and Voyager — no mission had contributed as much to understanding the planet as Cassini had. But was four years, the extent of the Cassini mission, enough to study Saturn? There was a mighty cry of "No!" from scientists around the world.

Talk of extending Cassini's mission was announced in February 2007. Although the mission was to end in July 2008, researchers realized the value of extending it another four years. They were excited to peer deeper into Titan, to uncover the mysteries of the vapors that were coming from Enceladus, and to better understand Saturn's rings and clouds.

After the extension to 2010, the mission was further extended to May 2017, just after the planet's summer solstice, and has been dubbed the Cassini Solstice Mission.

This false-color portrait of Saturn was taken using Cassini's visual and infrared spectrometer. Saturn's internal heat sets the planet on fire on its night side (right). The pronounced difference between the northern and southern hemispheres is also captured here — the north evidently much brighter than the south. The reason for such a difference is due to the high-level particles that exist on the planet, which are more prevalent at the southern hemisphere. This results in more light escaping from the north and making the north appear brighter in this image.

One of the things that undoubtedly makes Saturn such an appealing target is the beauty of its rings. But the soft hues of the planet also make it a warm and inviting sight to the human eye.

In this photo Cassini looks slightly upward at the giant, at an angle of about 19 degrees below the ring plane. Just peeking out below the planet's western limb (bottom left) is its moon Janus.

How many moons can you see? This beautiful photo, taken by Cassini in July 2008, just before the planet's spring equinox, was created using 30 images taken from the spacecraft when it was approximately 69 million miles (1.1 billion km) out. The mosaic shows Saturn in its natural-color view, and reveals three of the planet's moons: (from left) Titan, Mimas and Pandora. In a larger version of the image scientists were able to identify three additional moons: Janus, Epimetheus and Enceladus.

This mosaic (above) captures the delicate beauty of Saturn in its natural colors as it sits nestled in the sun's shadow. Cassini took 45 separate images in May 2007 while the planet was slowly easing toward its 2009 equinox.

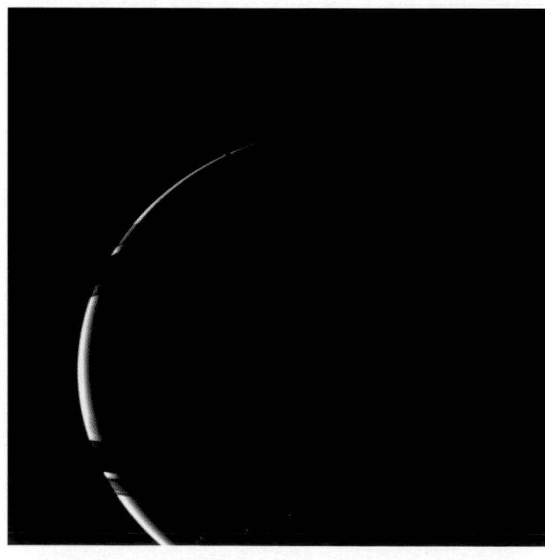

This dazzling display of Saturn's night side (left) shows the thin crescent of what is undoubtedly the most spectacular planet in our solar system. Cassini took this picture on July 31, 2006, at a distance of around 1.6 million miles (2.6 million km). The image shows the interaction of Saturn's shadow on what seems like a delicate ring of ice around the planet.

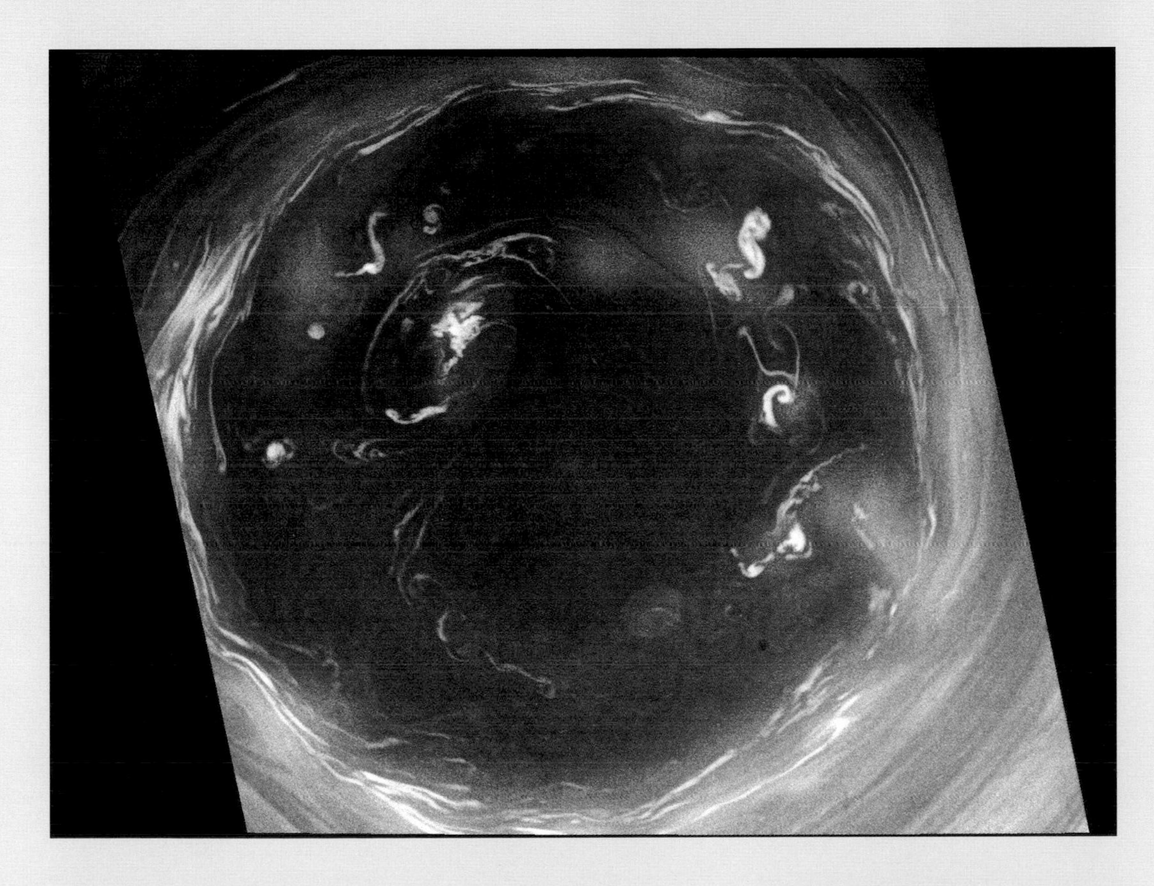

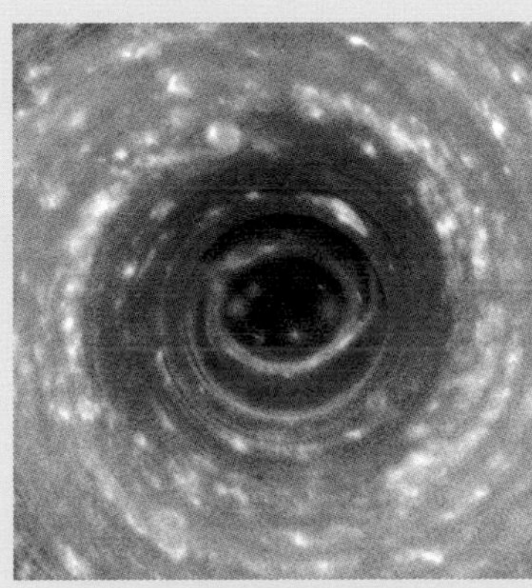

This tempest, located at Saturn's south pole, rages on in an incredible display of force. This phenomenon — of a raging storm, with such a well-defined vortex and eye structure — has never been seen on any other planet. Even Jupiter's Great Red Spot has no eye wall and is fairly calm at the center of the structure. And if you think that Earth's hurricanes produce terrifying wind speeds, this fierce, swirling mass has winds that whip around the planet at 342 mph (550 km per hour) and reach approximately 5000 miles (8,000 km) across. Also interesting to note is that this storm, unlike the hurricanes on our home planet, is relegated to Saturn's pole and does not drift. This image was taken on October 11, 2006, at a distance of 211,300 miles (340,000 km) from the planet.

Another look at the south polar storm shows shadows that allow scientists to uncover yet another bright ring of clouds. They believe that this is an indication of "convective upwelling," which helps them better learn how heat energy is moved around the planet's atmosphere.

Another high-resolution examination of Saturn's south-polar raging vortex helps researchers understand the convective activity that takes place to fuel such a powerful and seemingly never-ending storm. This image is 10 times more detailed than any other previous photographs and illustrates how much research is needed to fully comprehend the storm's structure. Other images had shown high clouds with a few "puffy" areas that circulated in the storm's center. However, this high-resolution image detailed that the "puffy" clouds weren't clear air, as previously believed, but rather other fierce storms that formed yet another inner ring in the larger storm. The ring is similar to a hurricane on Earth, but significantly larger.

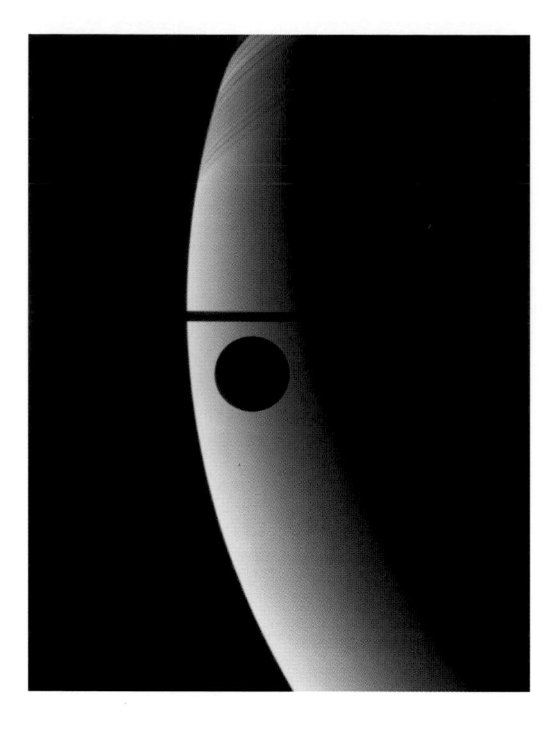

Dione sits atop Saturn's glorious rings as they cast their shadows across the surface of the planet. Saturn is a pale orb, full of delicate hues ranging from light beige to soft blue. But scientists are unsure why there is such a color variation from its south pole to its north. In the 1980s, the Voyager spacecraft had witnessed a more evenly colored planet. The blue color that Saturn has come to be known for was captured by Cassini in 2003, even as it approached its final destination. At that point, the planet was emerging from its northern hemispheric winter. But it's not just the colors that are evidently different on Saturn: the clouds vary at the poles and the equator. The polar regions are covered with sparse, bright clouds, while the middle of the planet has ever-moving cloud bands with the occasional vortex. Such differences are another reason why scientists aim to study the morphology of the planet and its enigmatic atmospheric conditions.

Sometimes the beauty of Saturn and its children is simply breathtaking. Although we know that scientists are taking part in this mission to extend their knowledge — and ours — of the planet, they study astronomy because of a deep passion and appreciation of our galactic neighborhood. Although we're currently mostly confined to the bonds of Earth, how long will it be before we can voyage to the depths of our solar system and take in the awesome sights that our ambassadors have been able to view for themselves? This image was part of a movie of Saturn's moon Rhea as it glided across the pale surface of the planet. The shadows of the rings are evident at the top of the image.

Things are looking up for Cassini. This image was snapped as it looked up from the planet's south pole. Such detailed, high-resolution images had never been taken before. The spacecraft was able to snap this photo on February 1, 2007, at a distance of a mere 584,000 miles (940,000 km). Cassini has been able to take unprecedented images from novel vantage points, making this mission undoubtedly the most successful planetary mission in terms of available study areas.

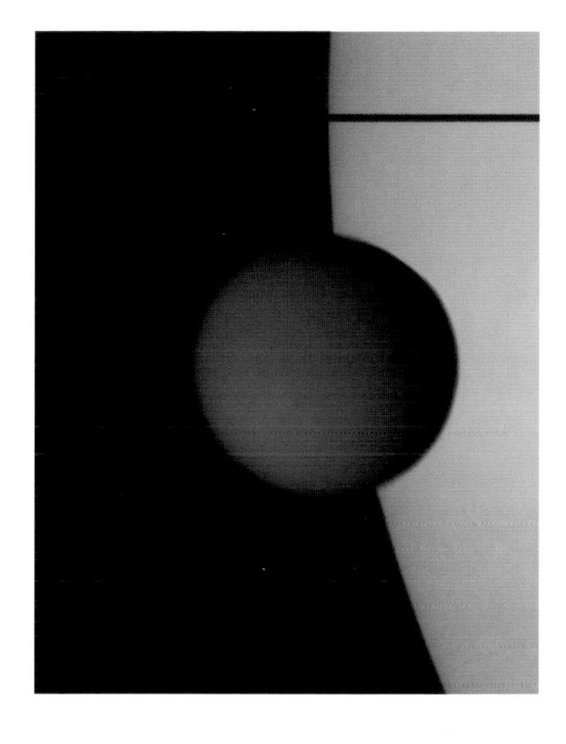

Ever since it was discovered that Titan had an atmosphere, the giant moon has intrigued researchers and astronomers. It stood to reason that, if it had an atmosphere, it could have water; if it had water and a weather system, it could harbor life. But Titan is not one to give up its secrets easily. The moon is covered in thick haze, making it almost impossible to see what lies beneath. The Cassini-Huygens mission was planned with that goal in mind — piercing the clouds to see what lies below the haze. This picture captured Titan as it transited Saturn. With the rings seen edge-on, and Titan in the foreground, this image serves to remind us about the delicate beauty that surrounds this system.

What makes Saturn such an appealing site for scientists and laymen alike is the allure of its sweeping rings. Such a seemingly delicate band around the giant planet begs more study. This photograph was taken while Cassini was 764,000 miles (1.23 million km) from the planet on June 19, 2007. The goal for researchers was to capture the rings in their entirety. In order to capture the full rings in their natural color, Cassini snapped several images with longer exposures, which resulted in the blinding white of Saturn's orb.

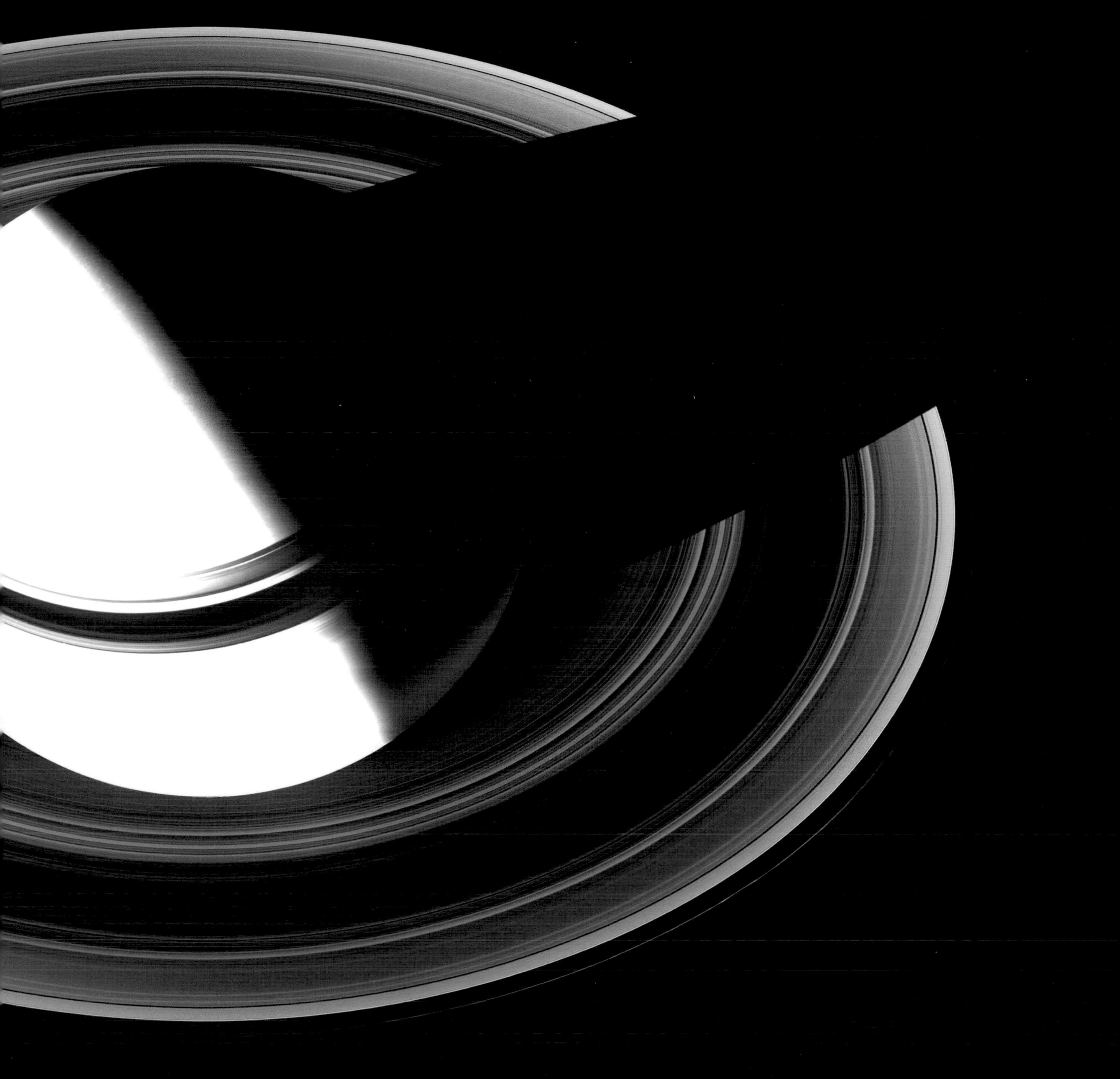

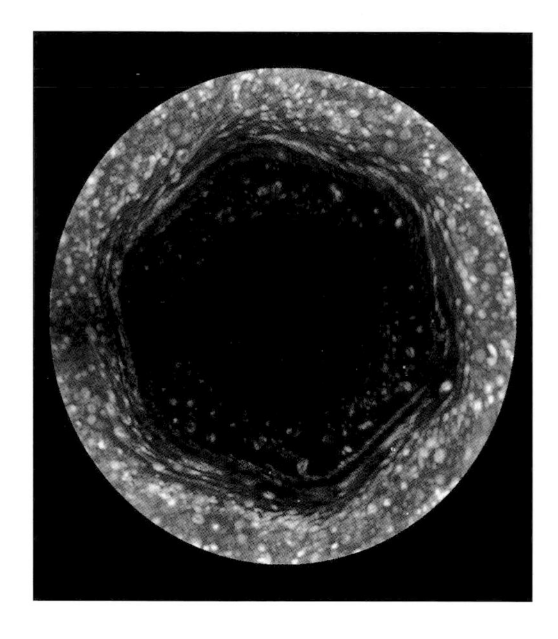

There are no straight lines in nature — but apparently there *are* hexagons.

This peculiar occurrence of a hexagon lies in Saturn's northern hemisphere. The image of this enigmatic feature was captured just as the planet's northern hemisphere emerged from its long winter — a winter equal to 29 Earth years. This was the first time that a complete picture of this phenomenon was photographed. Voyager discovered the hexagon in the 1980s, but was unable to photograph the whole northern pole. The hexagon remains a mystery; however, scientists believe that it is the result of a lingering jet stream. And even more questions have been raised, as the image also shows waves from the corners of the hexagon.

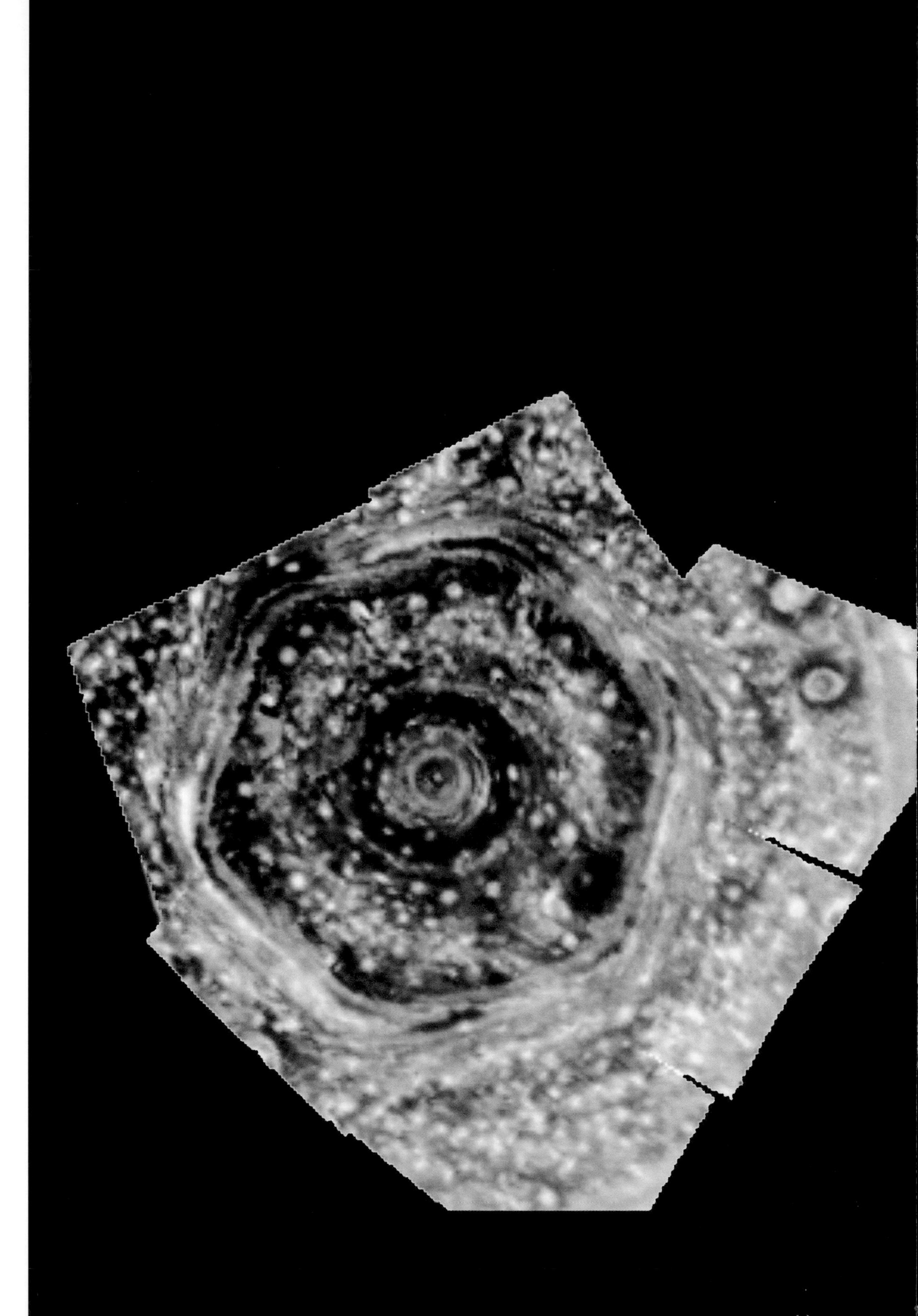

Saturn's complex atmosphere has created some fascinating phenomena. What is particularly interesting to scientists is the contrast in the storm systems that have developed at the planet's poles. The image on the left details the complex and poorly understood "hexagon" located in the north pole, with winds reaching 329 mph (530 km per hour). The image on the right is of the storm at the planet's south pole. The differences are obvious. Scientists believe that the small oblong clouds, evident at both poles, are convective upwelling and are responsible for powering both cyclones. The convective upwelling is thought to contain ammonia-hydrosulfide that lies below and is brought to higher altitudes. Both images were taken by Cassini's visual and infrared imaging spectrometer.

Not only the beauty, but also the extreme size of Saturn, is captured in this image taken by Cassini in 2005. Dione, tiny and isolated, floats silently as it crosses the solar system's second-largest planet. Dione is a moon with heavily cratered areas, mostly on the trailing hemisphere, which leads scientists to believe that a recent impact spun the moon around. Here it sits atop the rings, with its shadows cast elegantly across the pale orb.

When one thinks of Saturn, an image of pale beige hues instantly comes to mind. Images from the Hubble or Voyager probes have given rise to such a view. But Saturn also suffers from a bit of the blues. This true-color view of our solar system's friendly giant shows Mimas in the foreground of Saturn's northern hemisphere. At the time this image was snapped, the hemisphere was relatively cloud-free, allowing sunlight to take a shorter path to the planet, thereby scattering the light and giving the planet its bluish color.

Shadows of thin ringlets can be seen as Mimas crosses Saturn. Shadows of the rings are cast delicately as they fade into Saturn's night side.

Storm of the century? Perhaps. Cassini detected an intense storm in Saturn's southern hemisphere, nicknamed "Storm Alley," on November 27, 2007. Prior to this, scientists had noticed a relative lull in storms on the planet. Saturn experiences many storms, and Cassini had detected some of them much earlier in its examination of the gas giant. However, the earlier storms had lasted for less than 30 days, whereas this particular tempest raged on for several months. These images were taken three months after its initial detection. Scientists believe that an electrically charged storm penetrates the clouds from the planet's lower to upper troposphere.

This shows an earlier Cassini image, taken on December 6, 2007, of the storm that rages through the planet's southern hemisphere. Tethys can be seen on the lower left, with its shadow being cast across Saturn's northern hemisphere at the top left.

If we are going to escape the bonds of Earth and visit our planetary neighbors, we need to be prepared for the beauty that awaits us. This image of Saturn, composed of 165 composite images taken over nearly three hours on September 15, 2006, gives us an idea of what an approach to Saturn might look like. Not only was the satellite able to capture the astonishing beauty of the planet, but it also discovered never-before-seen rings. Learning more about Saturn and its rings will not only help astronomers understand how the planet formed, but it also gives them clues as to how our solar system developed, including Earth. From its position (which placed Saturn directly in between it and the Sun), Cassini also had the unique opportunity of sending back an image of Earth: the faint point of light in between the rings (to the upper left) is Earth, over 621 million miles (1 billion km) away.

Titan: A Strange New World

Titan is exactly what its name implies: a giant of the solar system. The moon is actually the second-largest satellite, closely following Jupiter's giant, Ganymede, measuring over 3,000 miles (5,000 km) across. It's only fitting that this enigmatic moon accompanies the solar system's second-largest planet in its dance through the cosmos. What makes this moon even more fascinating, and worthy of further study, is that it is very Earth-like in many respects. It possesses a dense atmosphere, with clouds that blow across the planet-like moon. It also contains mountains, rivers and lakes — not of water, but of methane. How are the lakes replenished? By the rains that fall from the clouds. Cassini's study of the moon has provided researchers with many answers, but also many more questions.

Scientists have long been puzzled by Titan. Christiaan Huygens discovered the first planetary satellite in 1659. More than 300 years later, G.P. Kuiper proved that the moon had an atmosphere — a surprising and exciting discovery. Since then, mainly through the discoveries made by Voyager 1, scientists have concluded that Titan has a rocky core and is made up mostly of nitrogen. But what's fascinating to scientists is that Titan is actually quite similar to Earth: it has an atmosphere containing mostly nitrogen, and even has weather patterns similar to our home. The planet also has lakes, but you wouldn't want to take a dip (even if you could sustain the surface pressure of 1½ times that of Earth's) — they're believed to be made up of mainly methane and ethane.

Not only does Saturn possess the rare and heavily defined ring system that it's known for, but it also has myriad moons with different characteristics. Here we see the small moon Tethys, at a mere 660 miles (1,062 km) across, silently slipping behind the massive Titan, which in contrast is 3,200 miles (5,150 km) across. The distance between the two in this image is about 621,000 miles (1 million km).

Titan's liquid methane lakes — theorized over 20 years ago — were finally proven to exist when Cassini's made its 2006 fly-by of the mighty moon. This colorized image (facing page) does not show exactly what we would see, but rather is a result of radar signals bouncing back to the spacecraft.

Unfortunately for people on Earth, the chances of anyone seeing the surface of Titan are pretty slim. But in this artist's concept (left), we are lucky enough to be standing in front of the Huygens probe, looking down at the ambassador from Earth on Titan's red and rocky surface. It is hoped that one day humans will be able to find a way to brave the elements of this moon to set foot upon its surface and explore this very Earth-like satellite.

The concept of visiting another heavenly body isn't new. Our ancestors long ago dreamed about how we could visit the stars, and even if we were to go back a mere 60 years, the idea of walking on another world was a fantasy. But then on July 22, 1968, most of the residents of Earth watched, mesmerized, as Neil Armstrong became the first human to ever set foot on another world (far right). Another 16 men followed in his footsteps. But what about other worlds? Will humanity one day be able to leave its impression on another dusty body? We would want to start off with a world a little more hospitable than Titan. The Huygens probe snapped this image (right) on the surface of Titan. Objects at the center of this image are roughly the size of a man's foot. Perhaps one day, an image similar to the moon landing image will be sent to Earth. But first, we'll have to find a way to endure Titan's harsh elements.

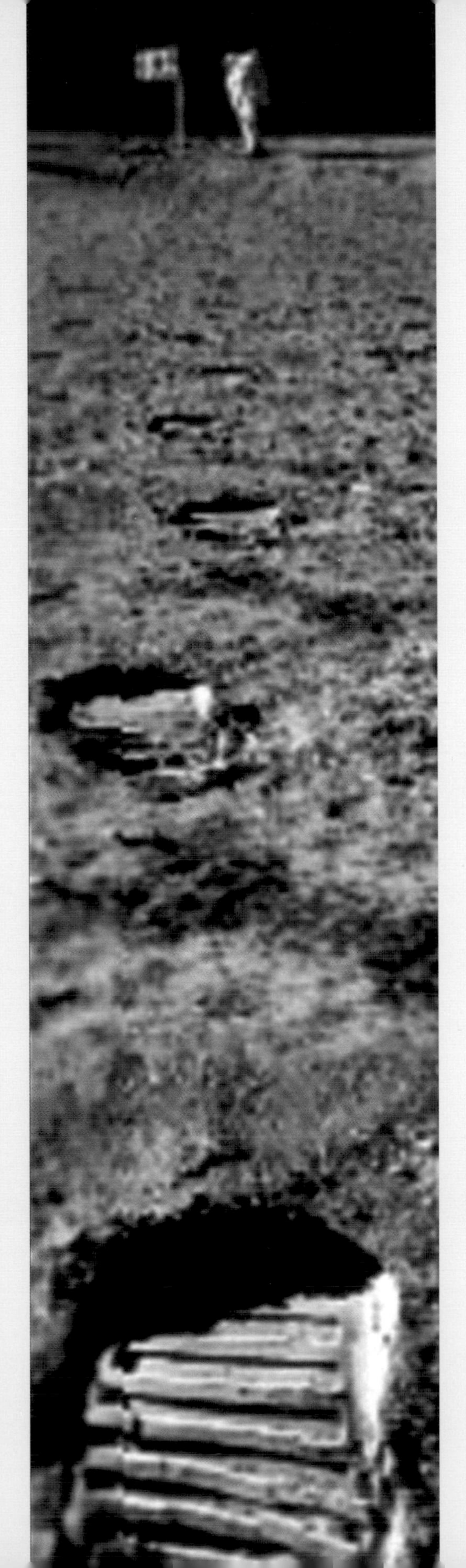

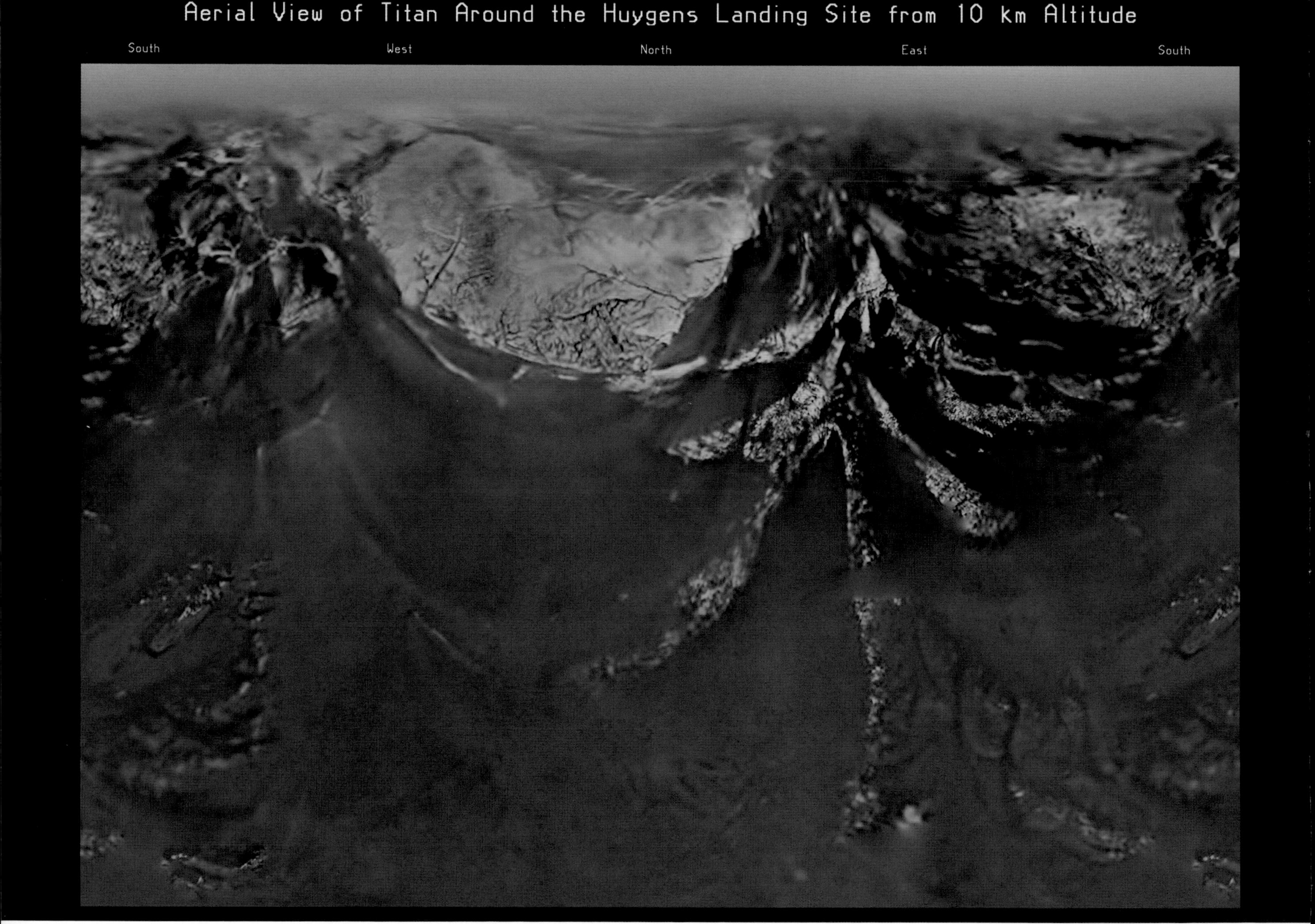

Aerial View of Titan Around the Huygens Landing Site from 10 km Altitude

South West North East South

Truth is stranger than fiction ... or in this case, the reality of what lies on other bodies throughout our solar system and beyond is perhaps stranger than we have ever imagined. Huygens snapped this picture of Titan's landscape. This is a Mercator projection map of the surface, taken while Huygens was gently coming to its final landing site on January 14, 2005. Never before had a spacecraft landed on such a fascinating and enigmatic body. The image to the right shows the individual frames the probe snapped as it glided to the surface.

Aerial Views of Titan Around the Huygens Landing Site

Altitude | West | North | East | South

150 km
100 mi

30 km
20 mi

8 km
5 mi

1.5 km
1 mi

0.3 km
0.2 mi

Want to go for a swim? Well, if you're visiting Titan in the near future, you might want to pass it up. Cassini's visit to our giant neighbor has revealed that Titan, the solar system's second-biggest moon to the solar system's second-biggest planet, is comprised of various lakes. But, unlike Earth's lakes, Titan's are made up of toxic (to us, at least) hydrocarbons. Like Earth's lakes, it is believed that Titan's are replenished by seasonal rains, but of hydrocarbons. The images on the left were taken on July 3, 2004, the ones on the right on June 6, 2005. The bright patches are clouds. The images reveal that, within this time span, large storms created the dark areas, believed to be lakes of liquid hydrocarbons.

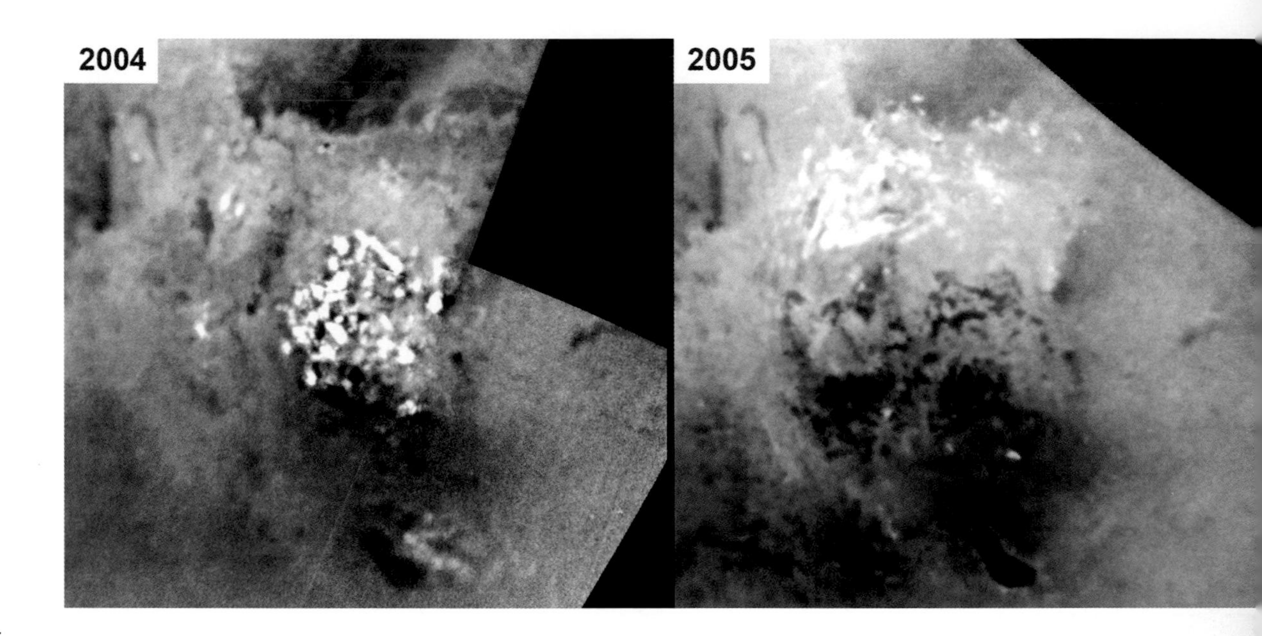

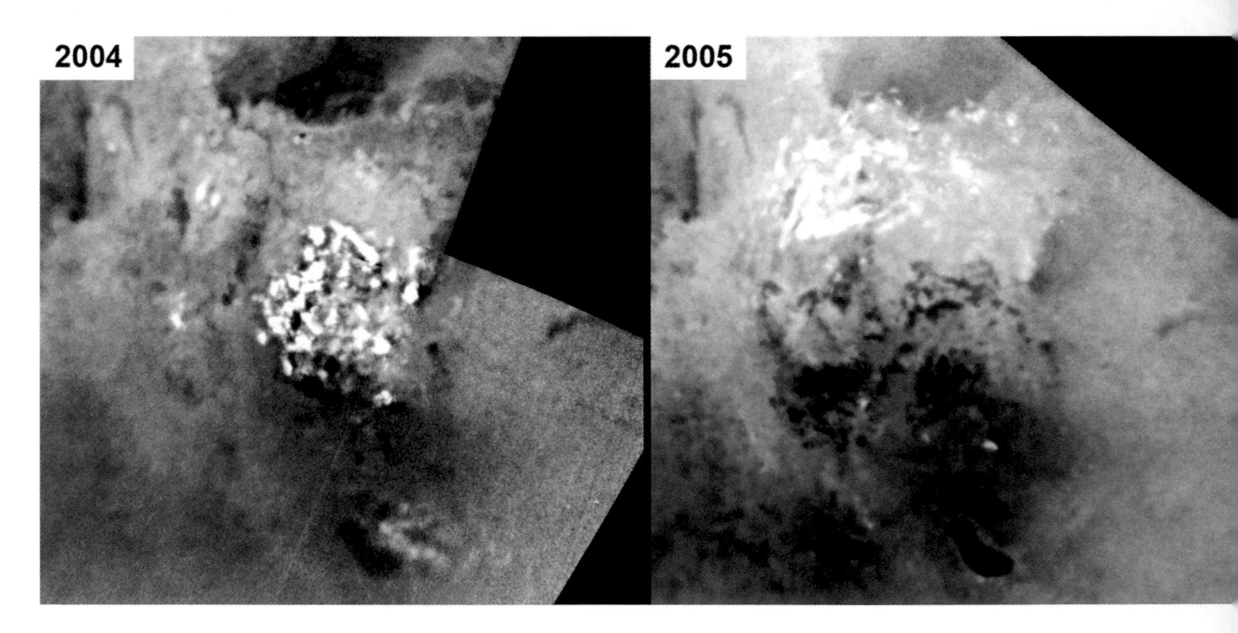

Here, Cassini acquired an almost perfect alignment of four of Saturn's moons: Dione, Titan, Prometheus and Telesto.

Titan eclipses the small moon, Epimetheus, in this portrait that details the prominent A and F rings of Saturn as they stretch across the giant smog-covered moon.

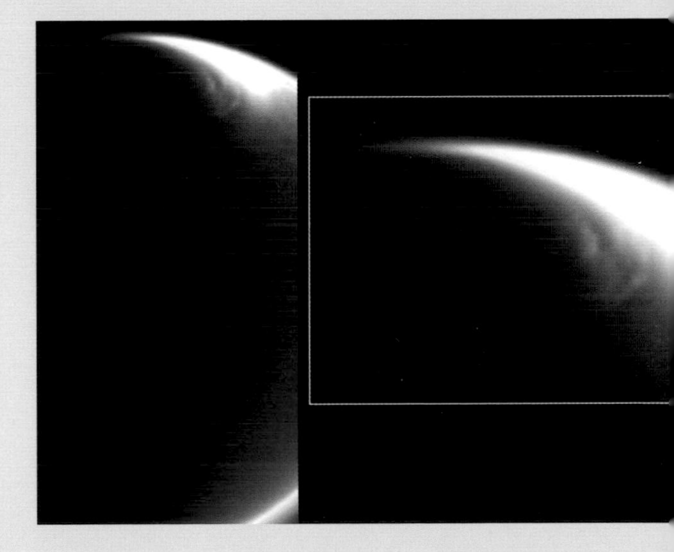

Scientists were eager for the Huygens probe to land and make observations on Titan. Were their predictions from observations and careful scientific analysis correct? Here, in a December 2006 fly-by of Titan, Cassini's infrared spectrometer photographed a spectacular cloud system that covered the moon's north pole. Although scientists had anticipated such a cloud, it had never been observed before in such detail. It also gave scientists further evidence for the moon's clouds being responsible for the replenishing of Titan's hydrocarbon lakes.

It was with great anticipation that scientists received some of the finest and closest images of Titan ever taken. This radar image taken by the Cassini-Huygens spacecraft is a smaller portrait of an image taken on October 26, 2004. The spacecraft flew just about 746 miles (1,200 km) above Titan's surface. These two images were provided to give a better understanding of what the moon could be comprised of. The colored image shows more detail, with brighter areas showing what could be rougher terrains and slopes. The pink areas show smaller details of the moon, with green areas representing smoother areas.

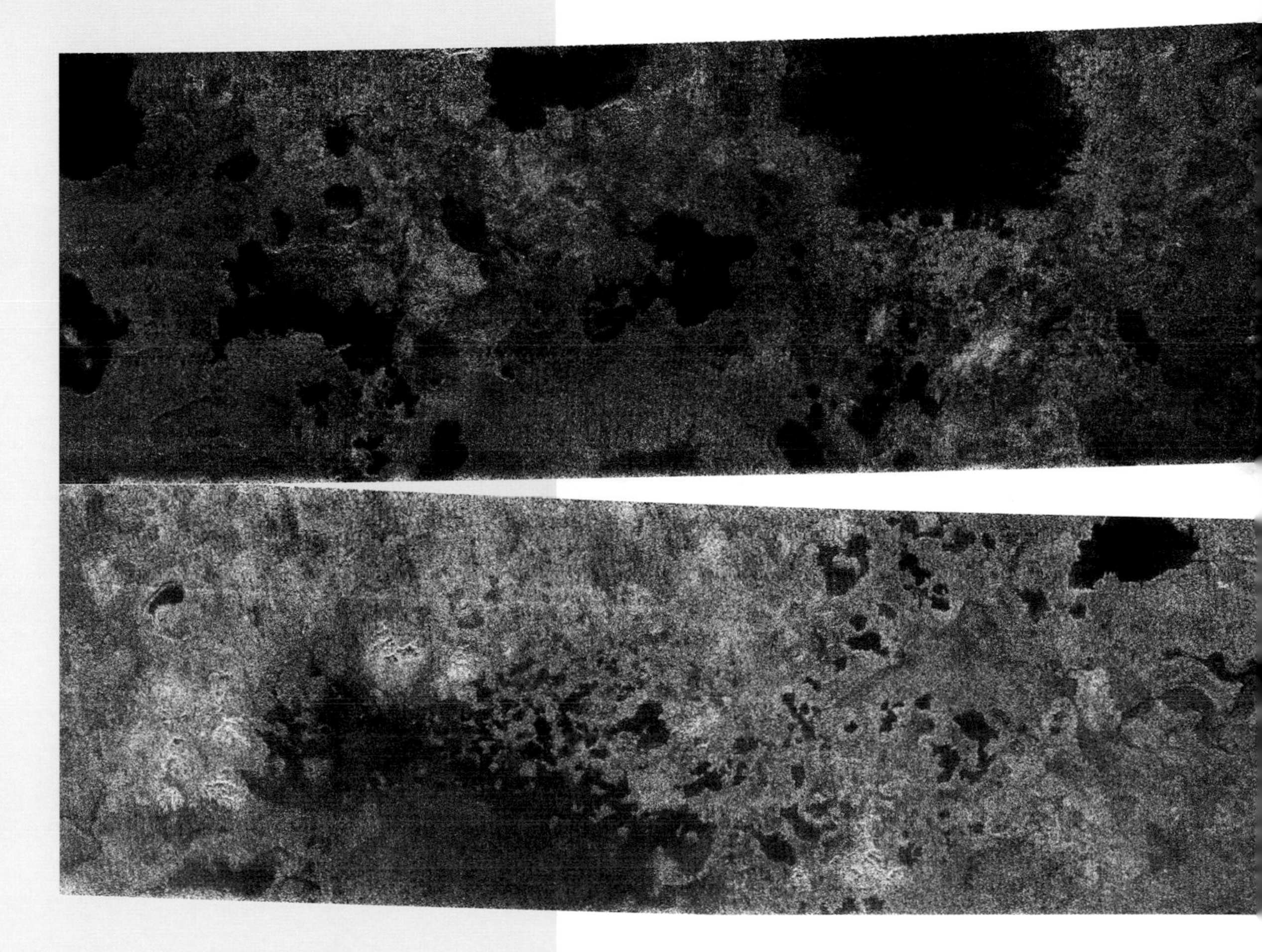

The Cassini-Huygens mission had one goal in mind: to uncover the mysteries of one of the most dynamic planetary systems within our celestial neighborhood. Among the many questions that researchers sought to answer was what lay beneath the hazy moon of Titan? It was theorized over 20 years ago that the body contained lakes of liquid methane. After years of study, scientists had theorized that Titan was replete with methane lakes, particularly near its polar regions. Using its radar system, Cassini photographed dark patches and channels. After studying the smooth, dark patches and what appeared to be rims surrounding them, scientists believed the dark patches to be lakes of liquid methane or ethane. This makes Titan the only other body in the solar system to possess lakes.

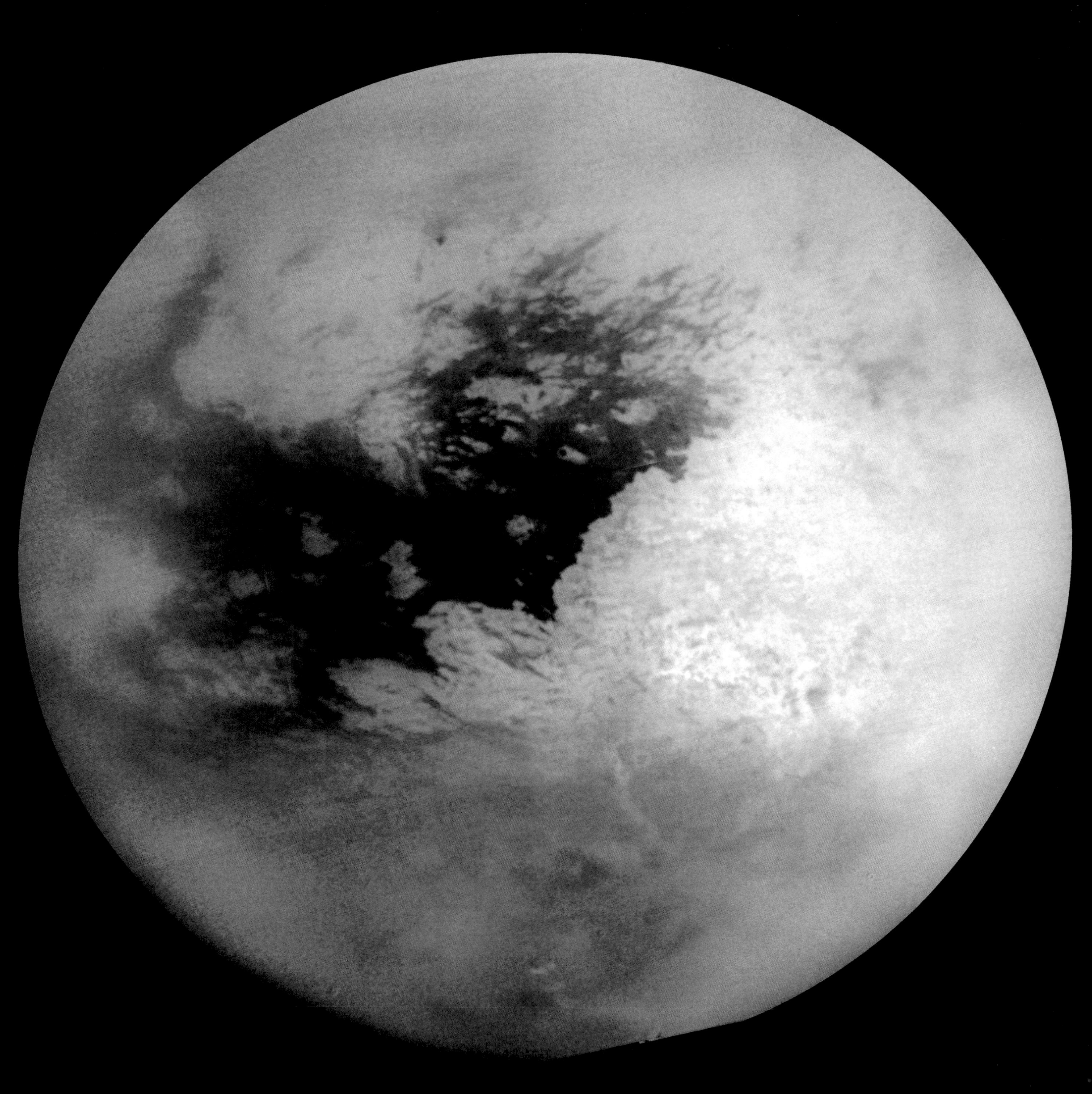

For so many years, Titan has been clouded in a shroud of mystery. But as Cassini flew past the massive moon, it managed to remove the cloudy haze that surrounded it, revealing actual features. This is a composite of 16 images that have had the effects of the moon's haze removed.

Saturn peeks out from behind the mighty Titan.

This image shows the largest lake on Titan (left) and compares it to the largest lake in North America — Lake Superior (right). The lake on Titan is believed to be at least 62,000 square miles (100,000 square km), greater than Lake Superior, which measures 51,000 square miles (82,000 square km). Proportionally, the Titan lake covers a larger area than our inland sea, the Black Sea. Because of this, scientists now refer to Titan's lake as a sea.

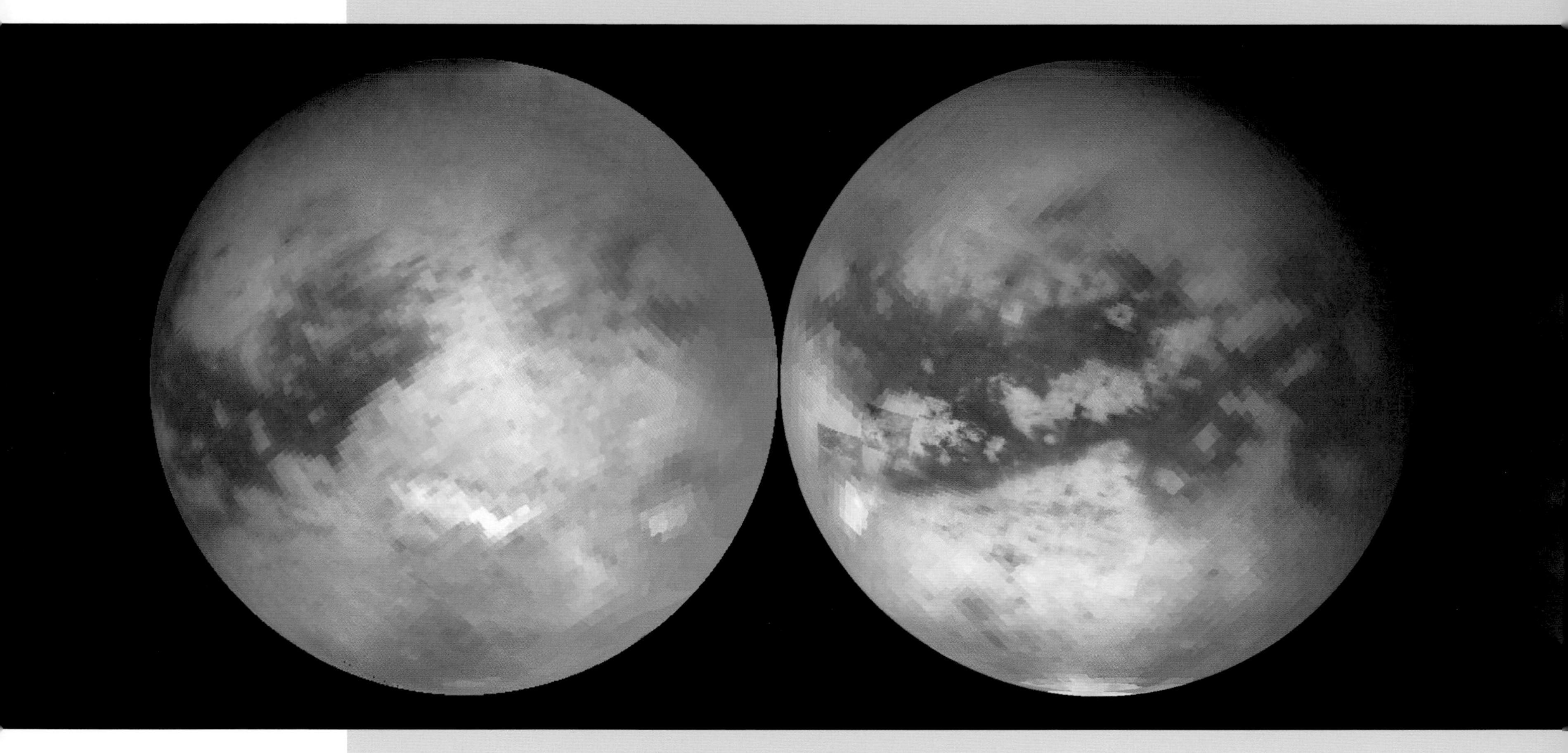

These three images were taken by Cassini's visual and infrared spectrometer, which allows the spacecraft to pierce the clouds of the moon to capture never-before-seen detail of Titan. The images were taken (from left to right) in October and December 2005, and January 2006. One of the most interesting features is Tui Reggio, which is believed to be a surface deposit of volcanic origin. Scientists believe that it could be comprised of water and/or carbon dioxide frozen from the uprising volcanic vapor.

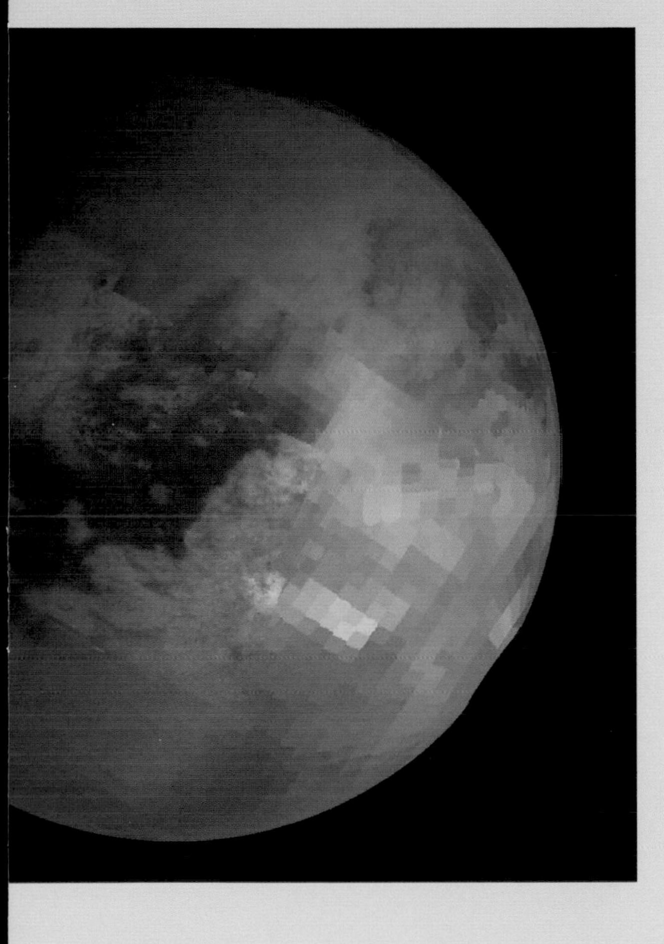

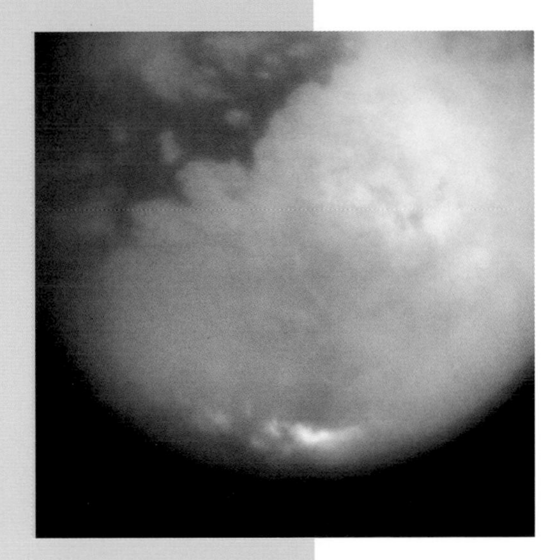

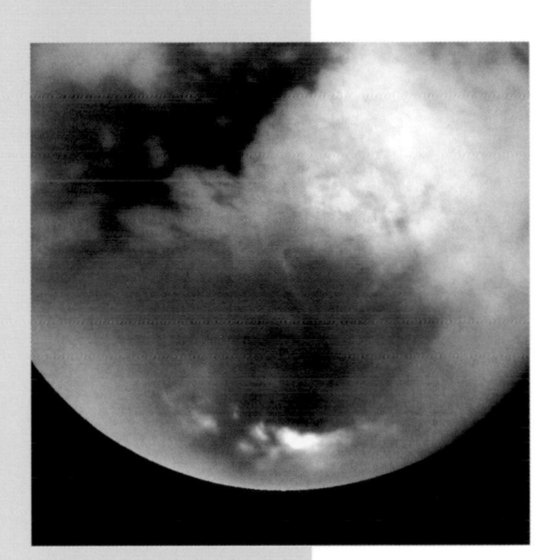

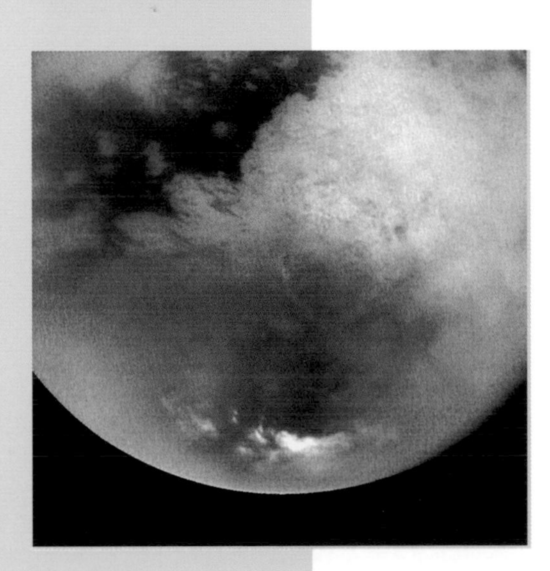

Cassini took this sequence of images when it was just 38 hours away from making its closest approach to Titan in October 2004. The images show how processing techniques can help scientists further define particular details on the moon. The first image has undergone a straightforward cleansing of noise and imperfections in the spacecraft's CCD camera. Multiple images have been used to create the second image, whereas the third image has been processed even more, further sharpening the moon's features. The bright area at the center of the images is Xanadu, a region that scientists had examined before using earlier images taken from Cassini and from Earth.

Majestic Rings

Since Galileo first peered up through his simple telescope to uncover the mysterious orb that wandered the night sky, we have been intrigued by what the ancient astronomer first mistook to be "ears" to Saturn. And ever since Christiaan Huygens uncovered the mystery of the rings, they have fascinated us, capturing the eye of both novice and professional astronomers. Part of Cassini's mission was to unravel the rings' mysteries. Some of the discoveries — as well as the photographs it returned — were stunning.

Saturn's rings are truly a thing of beauty. Believed to be created by collisions of moons or various bodies that were captured by the gas giant, they have formed a distinctive and amazing sight; knowing that each ring is comprised of billions of particles, scientists strive to better understand them.

This set of images (far right) shows just how closely they need to examine each ring to seek out small deviations that might expand their knowledge of the planet. Here, the B ring, Cassini Division and the F and A rings are all clearly evident. The middle image of the series is a closer examination of the area, approximately 1,118 miles (1,800 km). The two images farthest to the right show four small propeller-shaped features in the planet's A ring. This image was the first time that such small moonlets were observed in the planet's rings. Each moonlet, which lies within the propeller structure, is approximately the size of a football field.

Saturn's equinox presented scientists with the unique opportunity to study Saturn's rings in much more detail than ever before. With Cassini's high-resolution optics, and the sun being in just the right place, casting its light across Saturn, scientists were able to view the interaction of moonlets in its rings. Scientists believe that the shadow of the top image (shown in closer detail directly underneath), is the result of a moonlet that lies on an inclined plane to the rings, as evidenced by the shadow it casts.

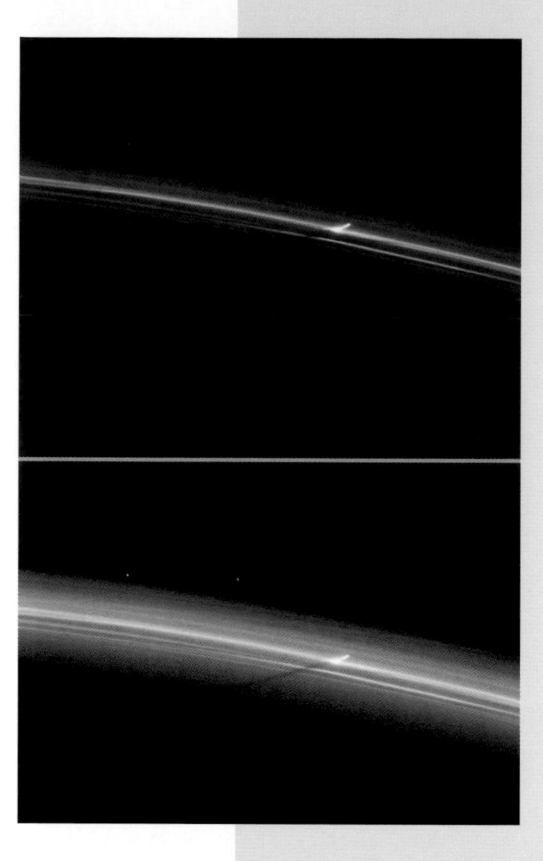

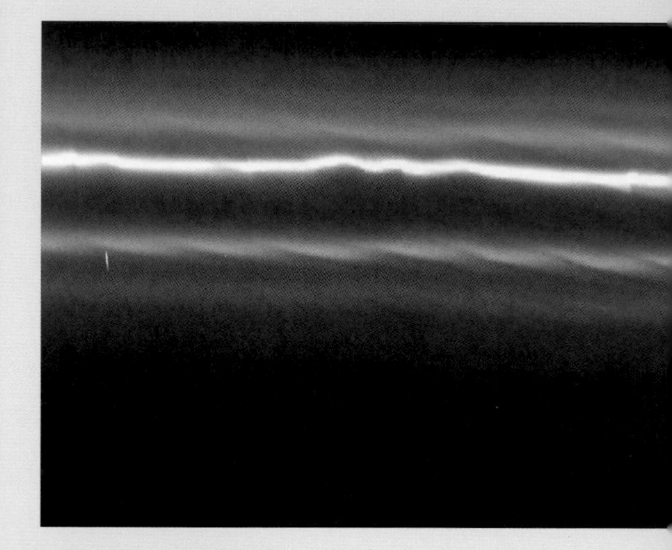

Saturn continues to present astronomers with pleasant surprises. These images, captured in just over 10 minutes, reveal a new moonlet in Saturn's faint G ring. Scientists examining the images believe the object to be reflected light from a tiny moonlet in the faint, dusty ring. The moonlet is believed to be only a third of a mile (half a kilometer) wide and is thought to responsible for the ring. Each image was taken in different light: the first in visible light, the second in red light, the third in near-infrared wavelengths.

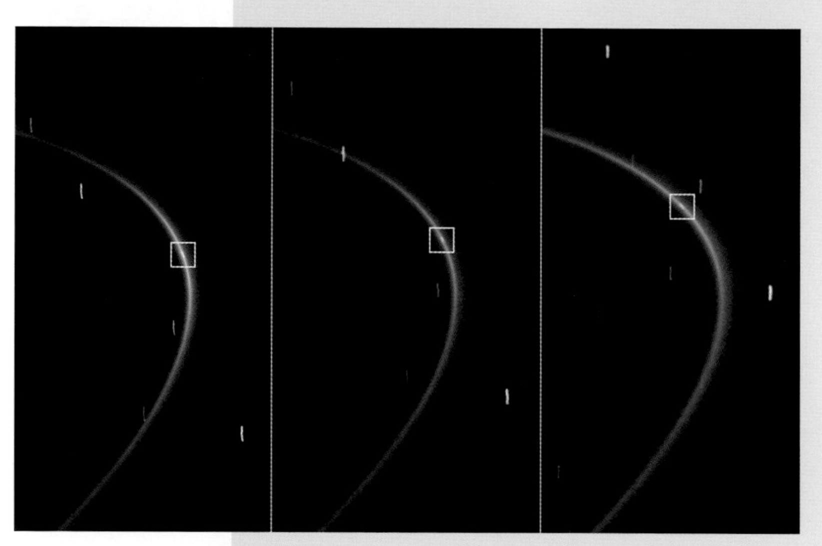

Prometheus is a relatively small moon, measuring only 53 miles (86 km) across, but it causes big waves. It has a very peculiar orbit that has an effect on Saturn's F ring, as evidenced here. As it approaches and recedes, it creates a "gore" in the ring system, once every 14.7 hours. A mosaic of 15 of Cassini's images has been compiled here to make the ring appear straight, and it shows a region approximately 932 miles (1,500 km) long. Prometheus does not actually enter the F ring; its gravitational pull, however, is responsible for waves. Each channel has a different tilt because the particles move more slowly than the moon, causing them to shear, and increasing their slopes over time. This effect has not been witnessed in any other planetary ring system.

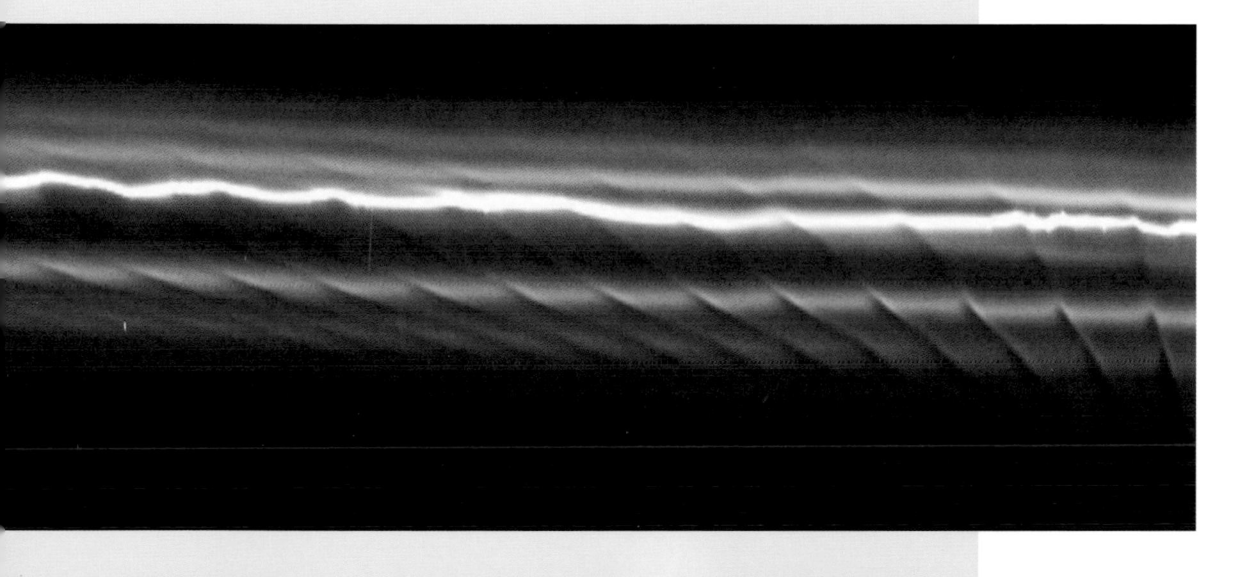

Saturn's shadow cuts across the rings as Prometheus sits at the far edge of the faint F ring.

Looking at this high-resolution image of Saturn's rings, it's difficult not to pity ancient astronomers like Christiaan Huygens, who only appreciated the simple beauty of the planet at over a billion kilometers away. Up close, we can see just how intricate the ring system is, with its stunning color. Its varying hues are quite evident. Although the color variations had been visible in images taken by the Voyager satellites and Hubble telescope, from this particular vantage point, the colors become much more prominent. The rings are thought to be comprised mostly of ice. Why the color variations? Scientists believe that the rings are also made up of other materials such as rock and hydrocarbons. The brightest of the rings is the large B ring. These images were taken nine days before Cassini took orbit around Saturn.

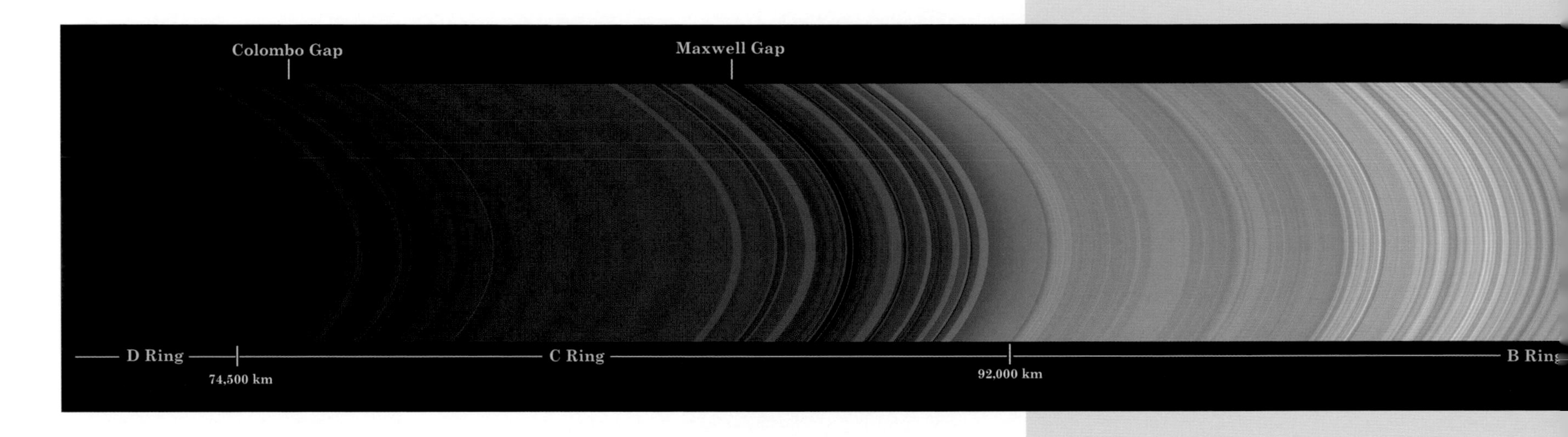

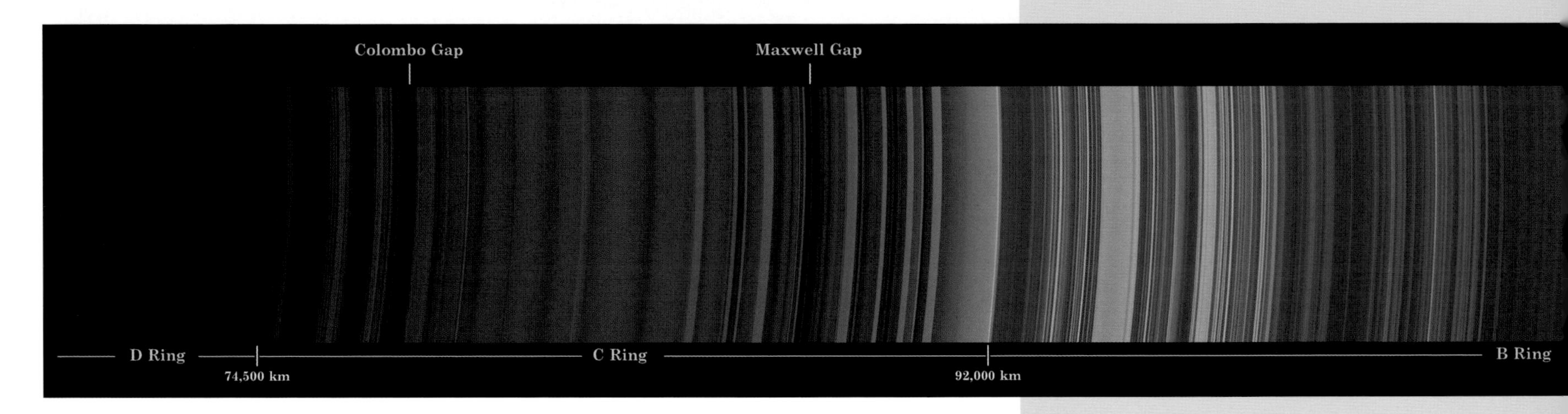

Ever since the discovery of Saturn's rings, we have dreamed about voyaging across the vast distances between our two planets to see for ourselves the beauty of the intricate system. As we have yet to perfect space travel, we are relegated to sending our robotic messengers to act as our eyes. And what spectacular images they have returned to us!

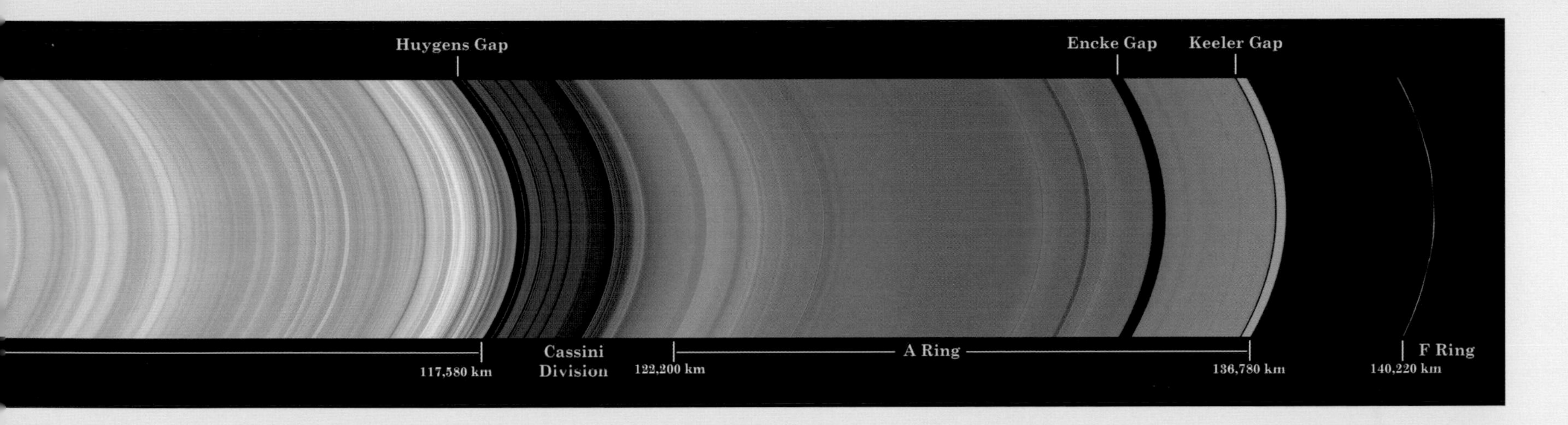

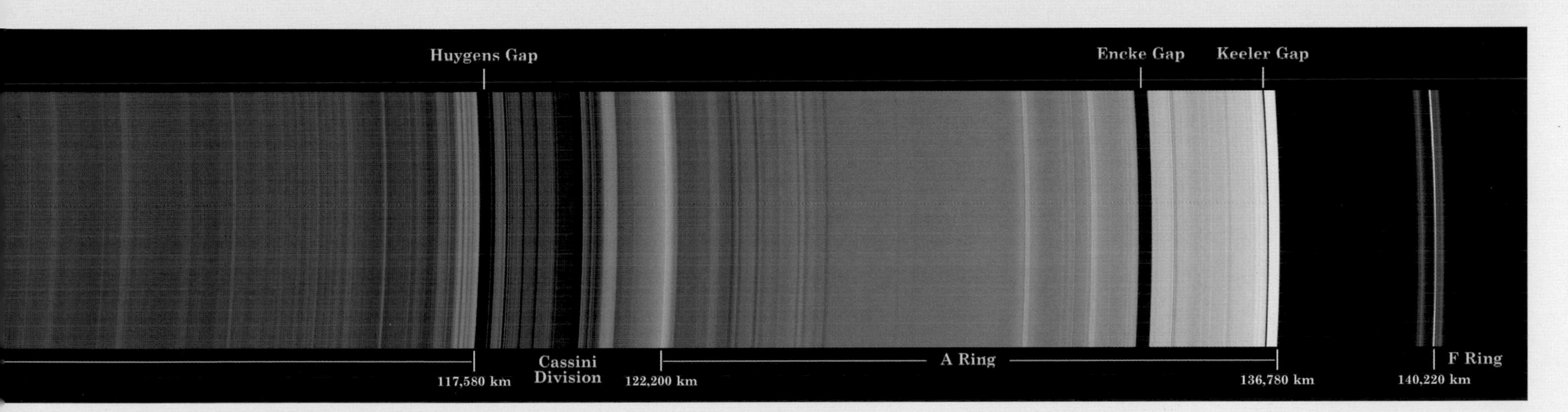

This natural-color detailed image of Saturn's rings (above) was snapped by Cassini on November 26, 2008, at 10 degrees above the illuminated side of the system. It is interesting to note the contrast of the color of the B ring between the illuminated side and the unilluminated side (below). When illuminated by the sun, the B ring appears brighter than the other rings; however, when seeing the unlit side of the system, the B ring appears darker. Scientists attribute that to the greater density of the B ring. The photo below was taken on May 9, 2007.

At first this might look like an old film image that's been overexposed. But this is the result of several images of Tethys crossing Saturn's ring plane, with the A ring on the left and the B ring on the right. Although the B ring is made of material that is much denser than that of the A ring, which doesn't allow as much light to pass through, the areas appear much brighter than expected. Scientists believe that this is a result of "Saturnshine" (our moon experiences the same affect from Earthshine). The Cassini spacecraft was able to capture these images as the planet approached its equinox.

Although this may look like just another image of Saturn's rings, this particular image has provided new insight into the complicated ring system that encircles the planet. As Saturn approached its equinox, Cassini's extended mission was able to provide scientists with stunning new revelations. Among the highlights: the D ring shows evidence that it is elevated above the ring plane; some of the gaps visible in the C ring show a brightness that indicates that they are thicker than was previously thought; and of course, the ever-interesting and complicated "spokes"

in the B ring are clearly visible. The shadow that stretches across the rings is from Dione (which is shown in many locations due to the time span over which the mosaic images were taken). Scientists were thrilled to obtain these images, as they're only possible around the time of equinox, which takes place about once every 15 years. Although the image may look grainy, there are stars shining through the rings, and the cosmic rays that struck the camera have created some noise.

The Cassini spacecraft was able to capture many interesting images during its extended voyage. Once again, the planet's equinox didn't disappoint scientists: Mimas cast its shadow across the ring plane in this amazing image. The months before and after the planet's equinox provided scientists with a once-in-a-lifetime opportunity to view such occurrences in great detail.

Saturn's rings contain myriad objects, as yet undiscovered by scientists, but eagerly anticipated by them. Once again, as equinox approached Saturn, it provided an unprecedented view of the rings and their objects. This new moonlet was actually detected by the shadow it cast across the B ring. The placement of the shadow tells scientists that it sits about 220 yards (200 meters) above the ring plane. Images like this are only possible around the time of Saturn's equinox.

When the Voyager spacecraft passed by Saturn in 1980 and 1981, it returned an image that has long since intrigued scientists: "spokes" in the planet's B ring. Cassini took the opportunity of snapping this photograph during the planet's equinox. This darkened the ring, allowing other structures to appear significantly brighter, giving researchers a deeper understanding of the out-of-plane interactions with the rings. The long shadow that stretches across the rings belongs to Mimas.

Although Saturn's ring system seems to be static, there are billions of tiny fragments of ice and rock floating through the system. Scientists were excited by this image returned by Cassini during Saturn's equinox, which showed evidence of the activity that they believe takes place. The bright streaks, highlighted in the above right image, confirm that a constant stream of interplanetary objects falls onto the rings. Researchers believe that these objects, traveling at tens of kilometers a second, are smashing into the rings, creating clouds of miniscule particles that have stretched out due to the orbital motion, forming the bright streaks captured. The image on the left shows an impact onto the A ring; the image on the right shows an impact on the C ring.

This portrait of the rings of Saturn was taken with the Cassini ultraviolet imaging spectrograph in November 2006. It provided scientists with a unique opportunity to witness star occultations (when a star's light is briefly overshadowed by an object, in this case, the rings), thereby allowing them to delve deeper into the composition of the rings. The occultations allow a determination of the amount of material contained within each ring. The most detailed view provided was of the B ring, which is considered to be densely packed with clumps of matter, interspersed with gaps that are almost empty. Another discovery that was made was that the ring material is organized and constantly colliding.

Stellar occultations occur when a star is eclipsed by an object such as a moon or planet. In this case, so far out in space, Cassini used Saturn's rings to serve as a measuring device, helping scientists determine the composition and density of the rings themselves (see previous image). This photograph captures just such an occultation. Here, the star's light passes through the dense B ring.

Curious Moons

Titan is a mighty moon, but it is not the only interesting satellite of Saturn; at the time of publication, over 60 moons have been discovered around the massive planet. With such an intricate ring system, more are constantly being uncovered. And each moon, in particular those small ones buried within the inner ring system, tugs and pulls on the rings, resulting in an interaction that is slowly being elucidated.

Saturn's moons vary in size and shape and even composition. Some, like Enceladus, are icy orbs. Others, like Phoebe, are rocky. And yet others, like Hyperion, are porous, mysterious bodies. Then there is the difference in size: the moons vary from as small as a few kilometers, to moons like Rhea, the planet's second-largest moon, at over 932 miles (1,500 km) in diameter. These differences are part of what makes Saturn so appealing to scientists: Saturn is like its own planetary system. There remain so many mysteries to unravel, and understanding Saturn's system helps us better understand the development of our own solar system, and therefore Earth.

Pages 72–73 show the surface of Dione, while Iapetus is pictured on page 73. To the right is an illustration of Saturn, its rings and just some of the largest moons found in its system.

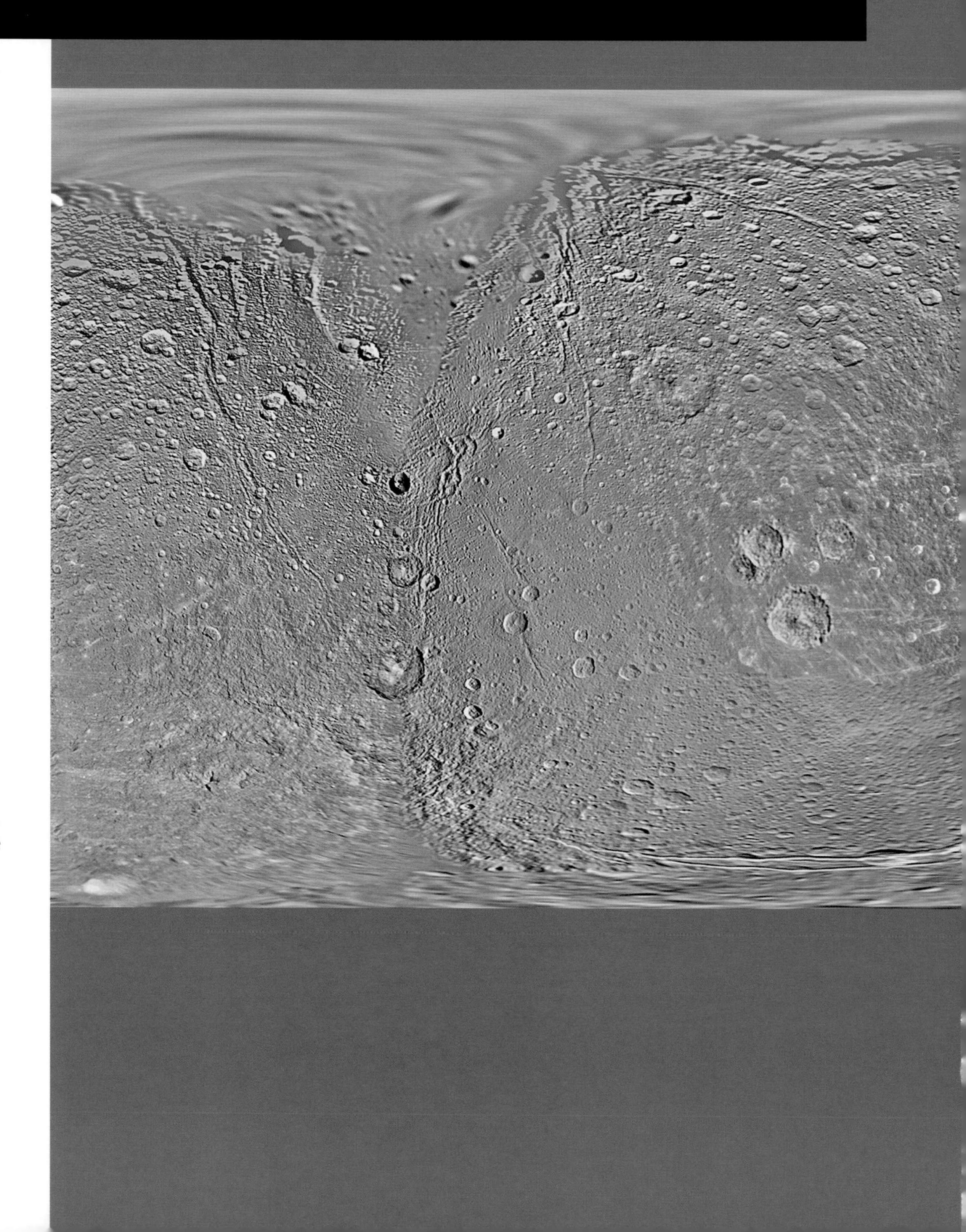

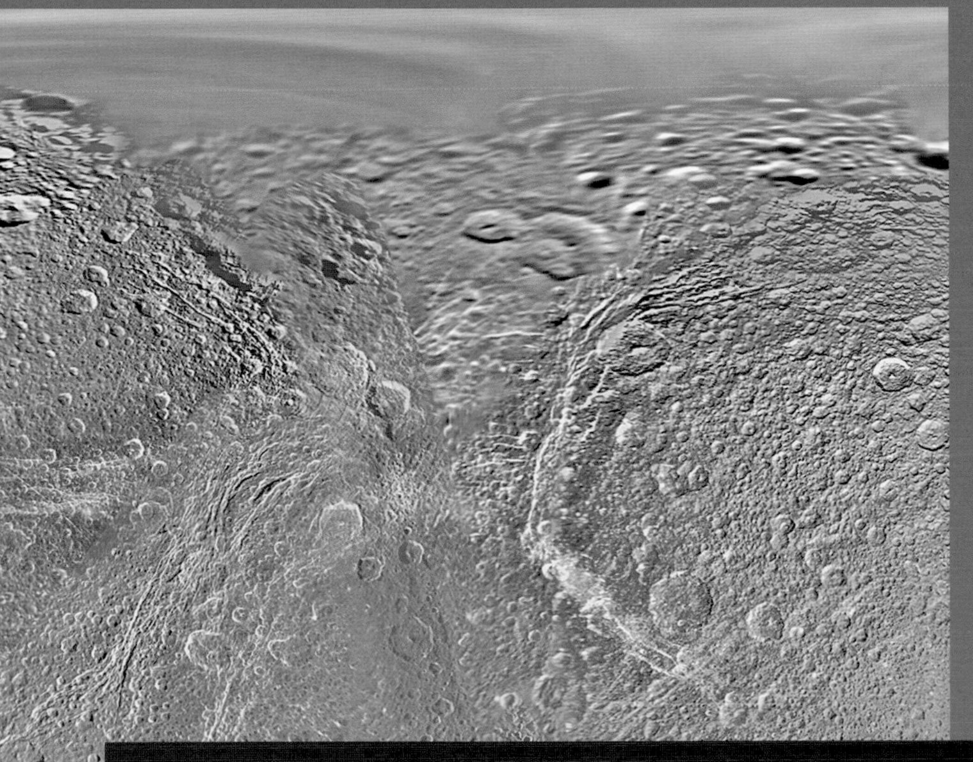

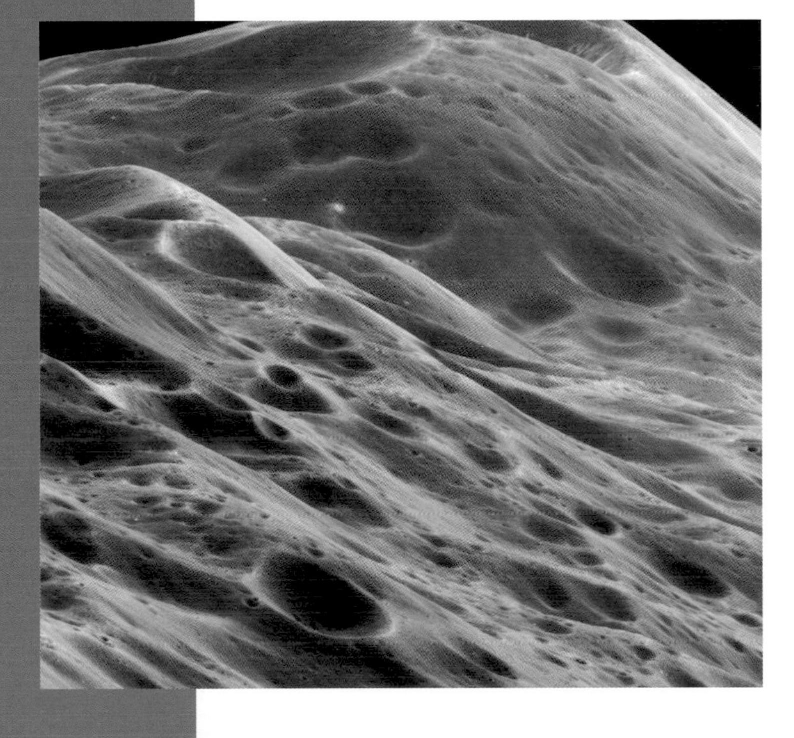

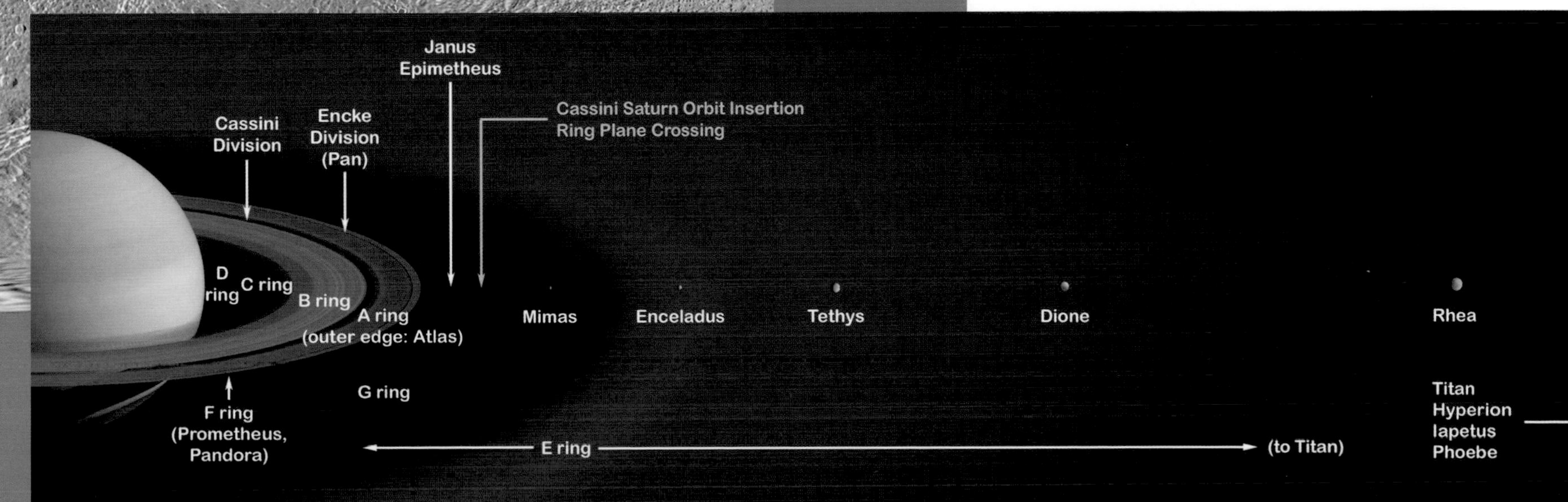
Janus
Epimetheus
Cassini Saturn Orbit Insertion
Ring Plane Crossing
Cassini
Division
Encke
Division
(Pan)
D
ring
C ring
B ring
A ring
(outer edge: Atlas)
Mimas
Enceladus
Tethys
Dione
Rhea
F ring
(Prometheus,
Pandora)
G ring
E ring
(to Titan)
Titan
Hyperion
Iapetus
Phoebe

Most people know that our moon has two faces. On Earth we are privy to only one side of it; our "man in the moon," with its craters and mare. The other, the "dark side of the moon," is heavily cratered. Other moons of the solar system have gone through similar evolutions. Iapetus itself has two faces. The color variations are quite evident. The image on the left shows the moon's leading hemisphere, while the image on the right shows the moon's trailing side. Both hemispheres have large impact craters. On the northern one (left-hand image, right side), the most well-preserved crater is called Turgi, measuring 360 miles (580 km) across. The right-hand image shows the other most well-preserved crater, called Engelier, which measures 313 miles (504 km) across. Why the color difference between the two hemispheres? Scientists believe that it is most likely due to thermal segregation of water ice between the two hemispheres. Usually, the differences are marked latitudinally, so why the difference on Iapetus longitudinally? In the study of the process on this moon, one model put forth is that red dust has deposited itself on the leading side of the moon, which is enough to make a difference. Here it evaporates the water on the leading side completely, but only marginally on the trailing side. Of course, there are other factors that may contribute to this, including its distance from the sun, its small size and surface gravity, as well as its outer position in the Saturnian moon system.

While many people on Earth rang in 2004 in various ways, Cassini rang in the new year in style as it flew past one of Saturn's many fascinating moons, Iapetus. The four images it captured were put together to show the many complicated geological structures of the moon. The dark, heavily cratered region, called Cassini Reggio, covers nearly an entire hemisphere of the small moon. An interesting discovery, among many that scientists found, is that the moon contains a topographic ridge that runs almost exactly along its equator, as evidenced in this composite image (above).

A rare view of the curious moon Enceladus was photographed by Cassini at just 16 miles (25 km) above the moon's surface. Astronomers are particularly interested in this moon, as there is evidence of geysers in at least 8 locations across its surface, located within these "tiger stripes." These stripes, which are fractures and folds, suggest tectonic activity, which is particularly interesting as Enceladus is a relatively small moon. The tiger stripes are called "sulci," one of the most prominent being Labtayt Sulci, which is located just above the center of this mosaic. Each feature on the moon is named after characters and places from *Arabian Nights*.

A virtual beacon of our solar system, Enceladus, shows scientists what it is made of ... mostly. In the 1980s, Voyager only uncovered a few craters on this icy and highly enigmatic moon. This led scientists to believe that some geological uprising was erasing the scars that should pockmark the ancient body. Then in 2005, Cassini descended to a mere 99 miles (160 km) above the moon's surface. The data that it returned showed that plumes of material were erupting from its south pole. Four months later, it collected images of eruptions of water vapor and ice particles. Researchers also discovered that the temperature at the pole, 100°F (38°C), was higher than they had expected. This meant that ice at the surface could melt. What excited scientists more, though, was the discovery of simple carbons. Does this mean that Enceladus harbors life? These discoveries just prove that further study of the moon — and Saturn's solar-system-like neighborhood — is needed. These images were captured by Cassini on November 21, 2009. More than 30 jets of vapor can be seen, with over 20 previously undiscovered.

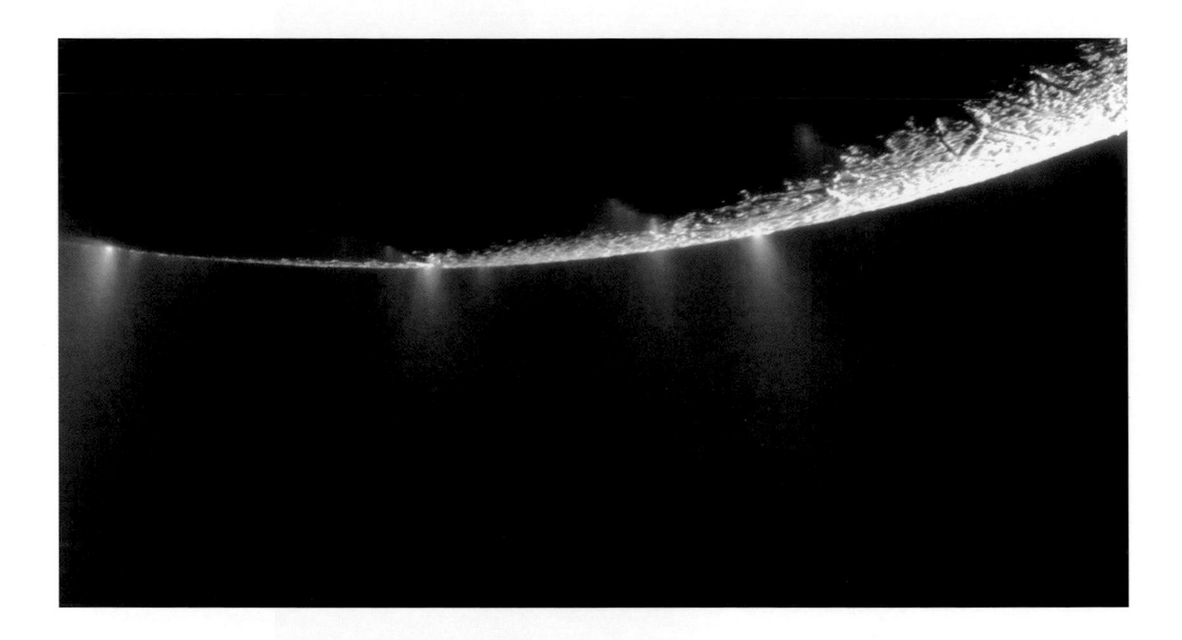

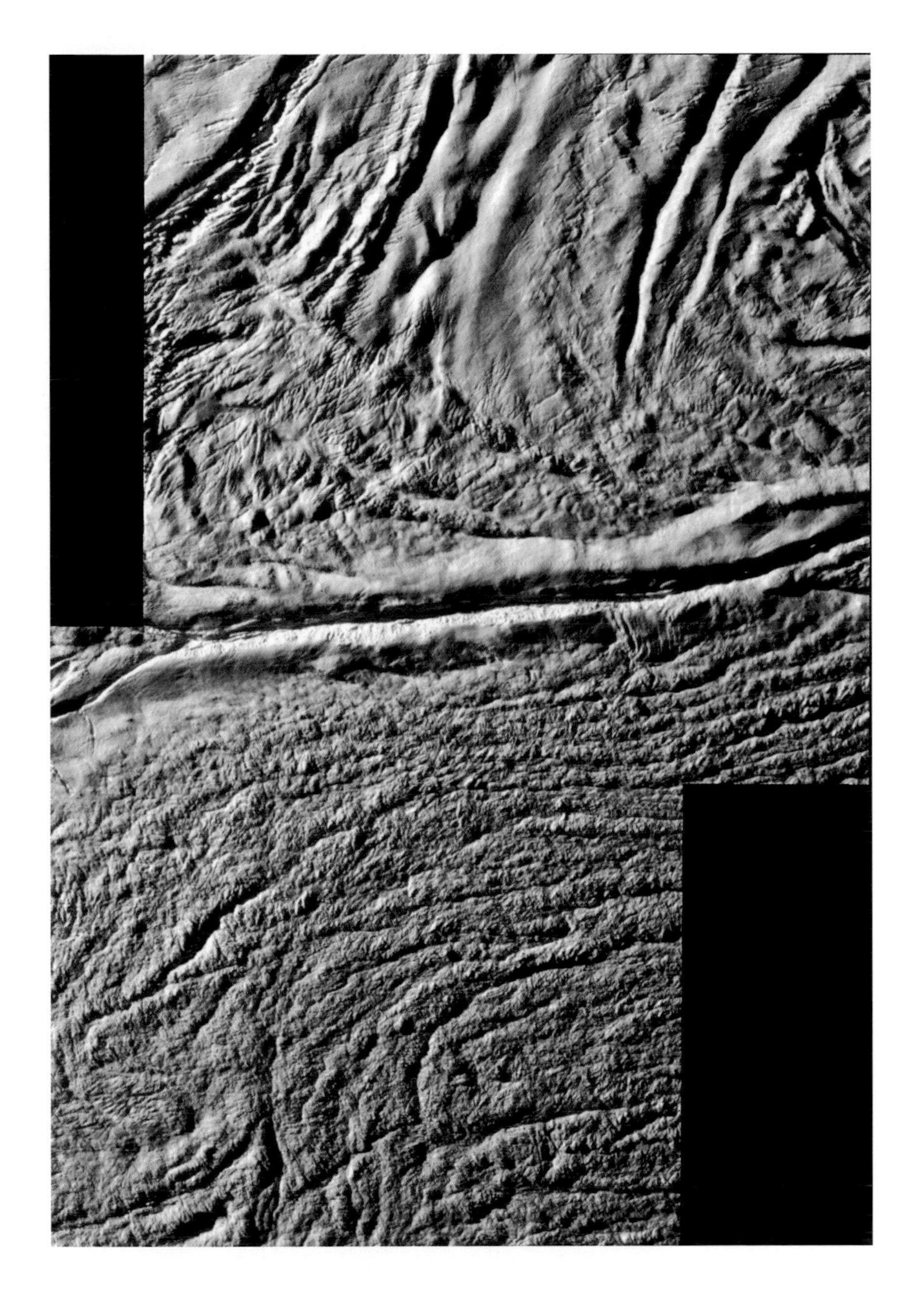

Enceladus's tiger stripes, or sulci, are uncovered in this high-resolution picture from Cassini as it orbited the moon in 2008. A composite of seven different images taken provided scientists with a deeper understanding of how these fissures in the moon's crust could be contributing to the eruption of water vapor and particles into space. Researchers are striving to determine the source of the eruptions, and theorize that it may be a result either of radioactive elements that are trapped inside the moon or to Saturn's squeezing of the small moon. Enceladus represents a never-ending puzzle, as it consists of at least five different types of terrain, including plains, craters and sulci.

Four high-resolution images have been combined to provide us with a spectacular view of Enceladus. Cassini flew by the icy moon on February 2005, and provided researchers and enthusiasts alike with an image that shows about 186 miles (300 km) across. It details the many fractures, folds and craters that have combined to make this moon such an intriguing target of study. When the Voyager satellites had flown by in the 1980s, they were able to provide scientists with various tectonic features, but nothing had been quite as revealing as this close fly-by. Voyager showed multiple terrains, but nothing like Cassini revealed in this high-resolution image, at almost 10 times the resolution. Scientists will now continue to attempt to ascertain the way this moon has developed these various fractures and features.

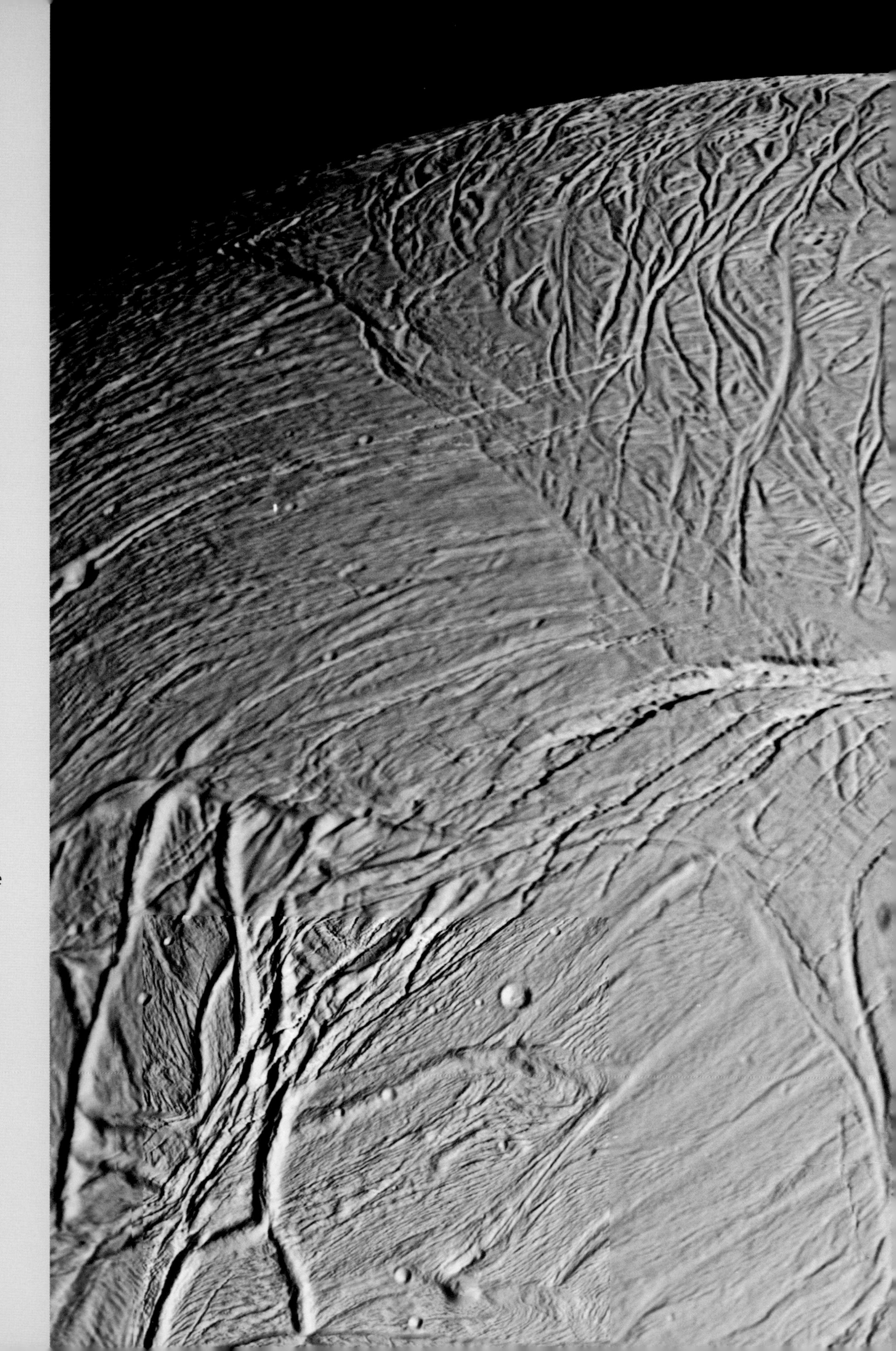

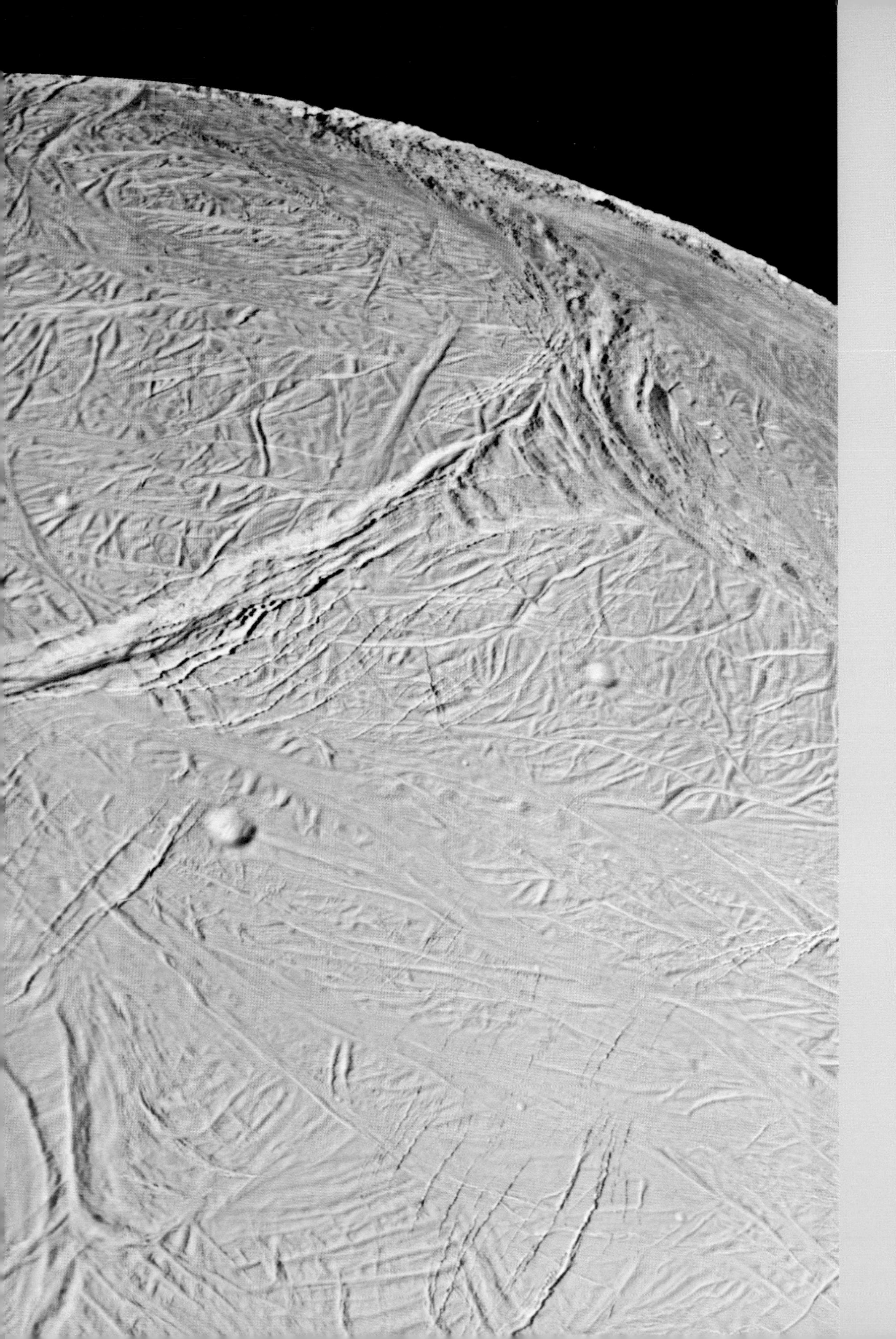

Unlike the image of the tiger stripes, or sulci, the other side of Enceladus tells a different story. This is the highest resolution image ever taken of the small moon's north polar region. What it reveals are heavily cratered plains, indicating an older region that was captured at different periods of disruption and changes in tectonic activity, most likely as a result of heating below the moon's surface. It also suggests that the icy crust was actually warmer at some period. The Voyager spacecraft did reveal this in the 1980s, but this image provides further proof and a much closer and superior view of the surface, which will enable scientists to compare the northern and southern polar regions' geological history.

When one thinks of a moon, invariably a picture is conjured of a sphere much like our own moon: a perfect orb, with mare, or "seas," and craters. But Saturn has captured many different types of satellites in its 4.7-million-year history. Here, Cassini photographed two sides of the small outer moon of Phoebe. The moon seems to have been battered by objects, some of which scientists believe to be smaller than 109 yards (100 meters). It is unknown whether or not the projectiles were from small asteroids or were a result of the formation of the planet itself. The crater to the right is evidence of a possible collision on the icy body that forced water up to the surface. High-resolution images like these help scientists understand more about the formation of Saturn's moons, and also about the planet itself.

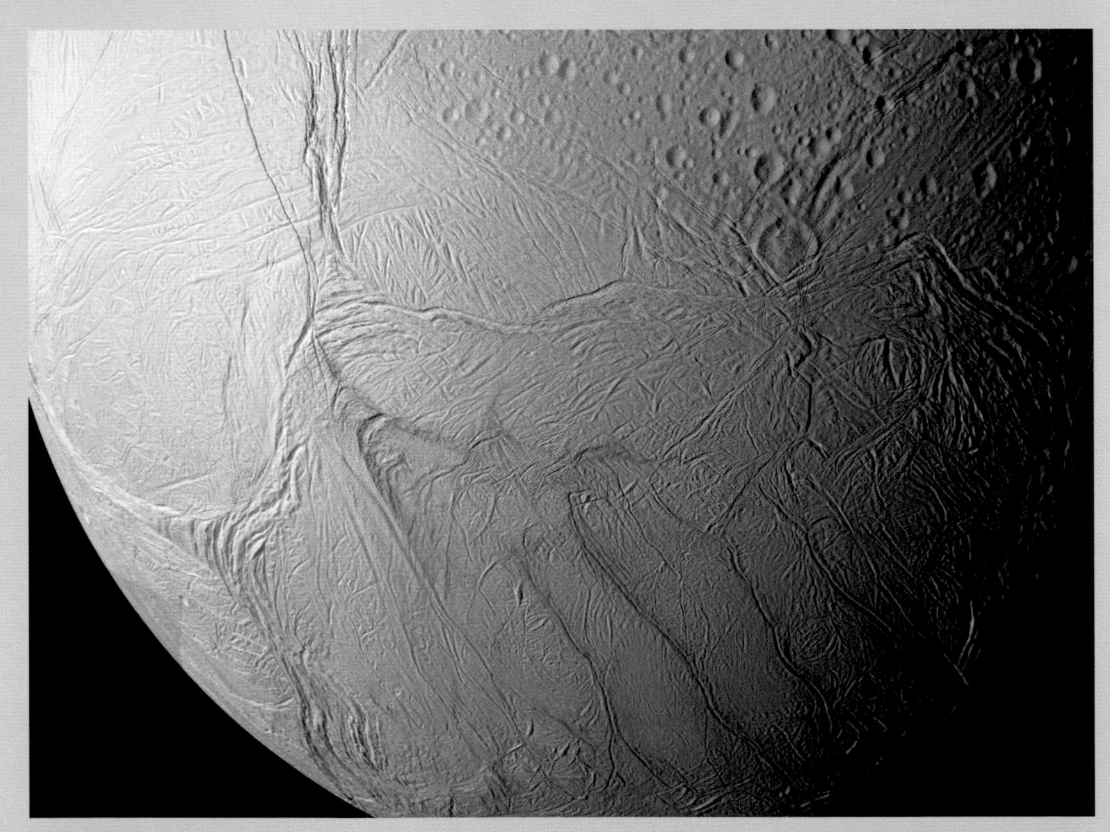
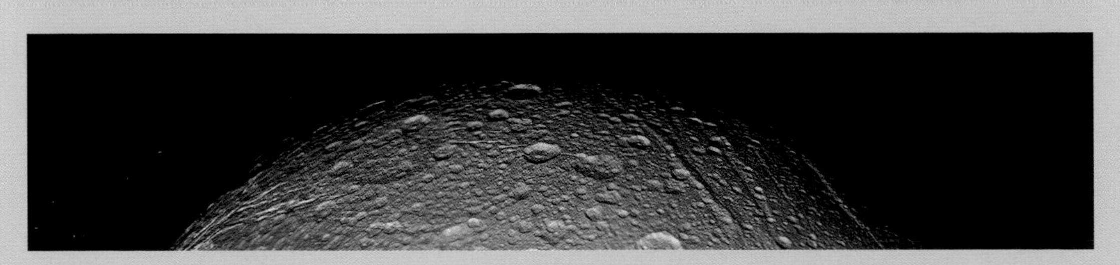

This stunning — and revealing — montage of four of Saturn's moons displays just how different many of the planet's satellites are. Each moon is made of ice, but has intensely different features. Each moon also shows impacts from meteors. But since scientists believe that each of Saturn's moons developed at roughly the same time, close examination of each individual moon helps them understand their various geological histories. Rhea (top left) and Iapetus (top right) have many impacts, which suggests to scientists that they have been exposed to impacts for eons. Dione's surface (bottom) suggests a younger moon. It also suggests that tectonic activity took place after the moon formed. As for Enceladus (center), its south pole points to more recent activity; its impacts are almost non-existent. And the discovery of a plume of material jetting out into space from its south pole confirmed that the region is still active. However, scientists continue to examine the moon closely in order to understand exactly what it is that is powering the activity on this moon.

Imagine not only being able to see incredible images of previously unexamined worlds, and being able to see the intense detail offered up, but also being able to name them. In this image of Phoebe's two sides, taken by Cassini in 2004, various features of the moon are named. The names come from the Argonauts, who were Greek explorers seeking the golden fleece. The largest feature on the moon was named after Jason, captain of the ship Argo. Phoebe is a relatively small moon, measuring only 133 miles (214 km) across.

One of the many things that excited scientists with the Cassini mission was being able to photograph areas on Saturn's various moons that had previously been unseen. Here, the moon Tethys is seen in its full glory, snapped by Cassini at almost 932 miles (1,500 km) above its surface. When Voyager passed by the moon in the 1980s, the high southern latitudes remained unphotographed. Yet here, scientists were pleased to get a glimpse of this region. One of the most evident features on the moon is Ithaca Chasma, the large rift that lies in the center of the image. The moon is also pockmarked, indicating a long history of impacts. What remains to be determined is why the craters have unusually bright floors. One theory is that the impacts hit through a brighter layer that lay underneath.

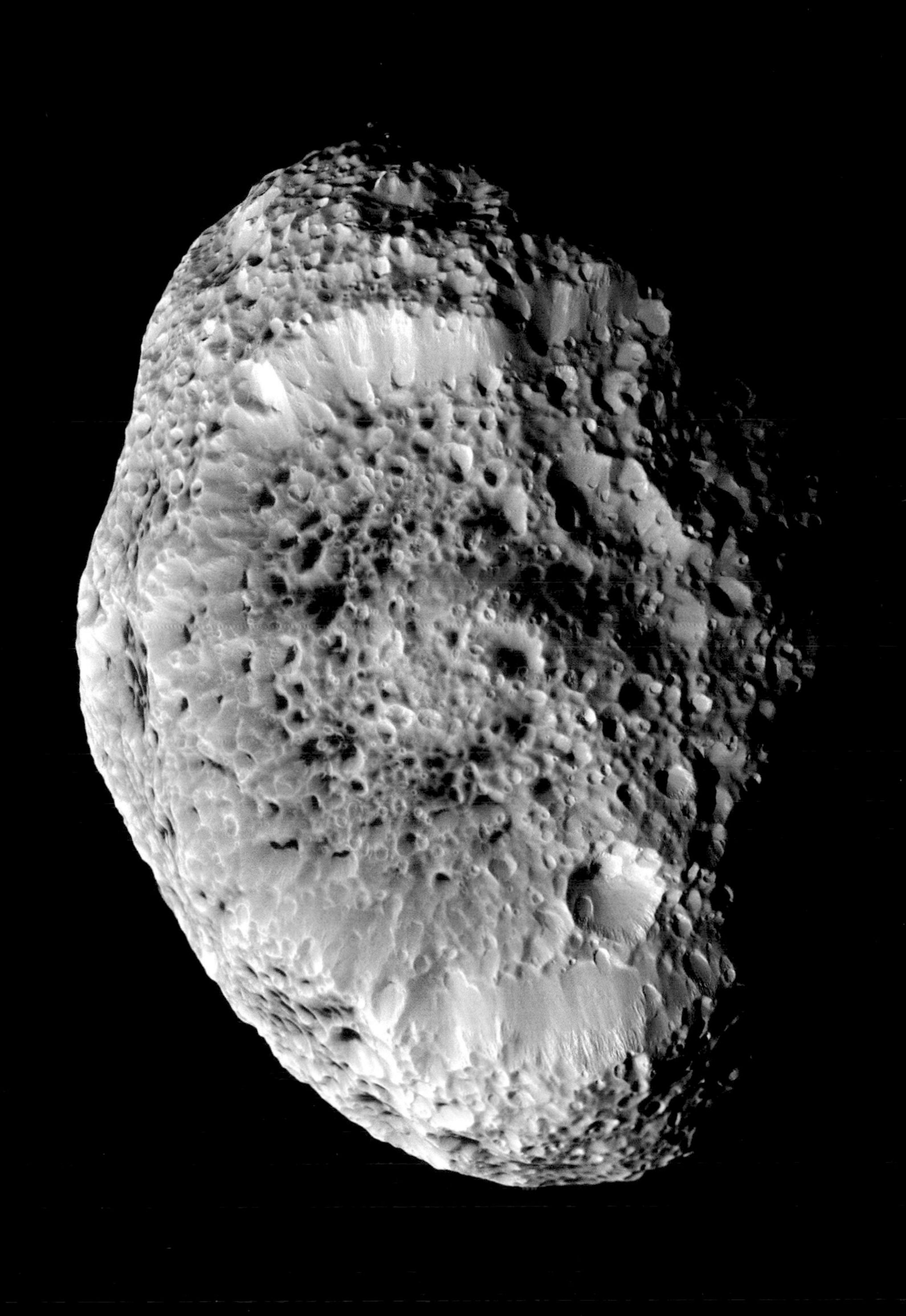

A giant potato or a giant sponge? This is Hyperion, one of Saturn's many peculiar and vastly different moons. It is also one of the largest-known irregular bodies in our solar system. Like Phoebe and Iapetus, Hyperion shows a heavily cratered surface, the result of its position as one of the farthest moons from Saturn's giant protective arms. Why the strange appearance? Scientists believe that the moon has a low density, which gives it very low gravity as well as high porosity. This contributes to the moon's appearance because it limits the amount of ejecta (what is brought up from the surface from an impact) that is left to coat the moon's surface. What little ejecta is produced is able to float away.

Like a delicate scale, Saturn balances tiny Dione in its soft hands. In this shot Cassini was on its way to the tiny moon, seeking to delve into the depths of a moon that seems innocuous and rather unspectacular.

It's interesting to note that, in natural color, Dione seems as colorless as it does in black-and-white images, while Saturn's delicate hues enliven the system. Dione is a moon that possesses heavily cratered areas, with some massive craters measuring 62 miles (100 km) across. Dione is 1½ times denser than water, which leads scientists to believe that the moon has a dense core.

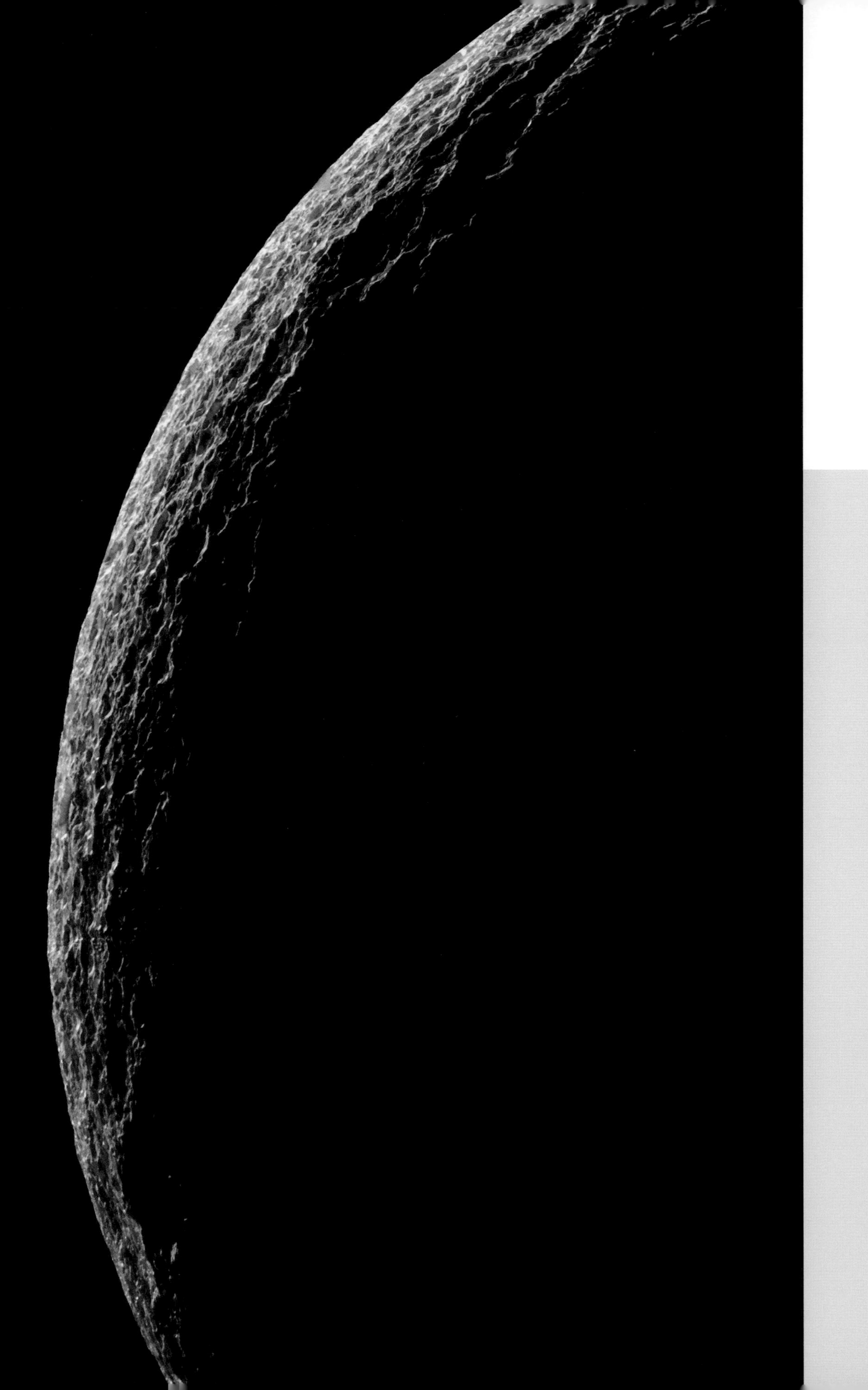

Tethys. The slim crescent of this cold and heavily cratered moon captures the beauty of such a desolate satellite. But it's not just a visually stunning photograph that researchers have taken: this image was an attempt by scientists to delve deeper into how the moon reflects light. Using this information, they can better understand what the satellite is comprised of. Ice and frost is reflected, with sunlight glinting off the craters.

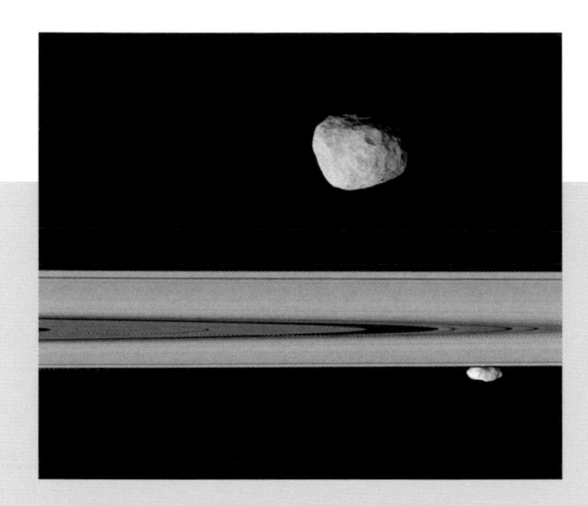

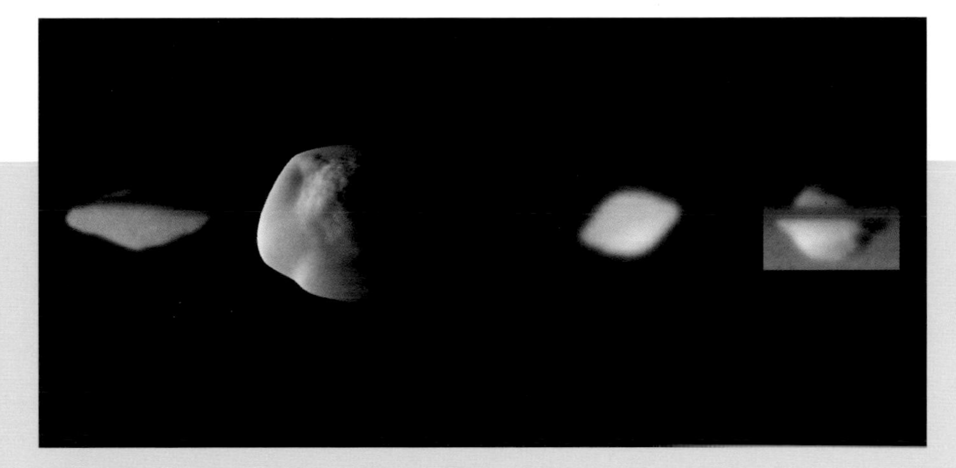

Here, Janus (above) and Prometheus (below) balance delicately in space on Saturn's teeter-totter rings. Janus is believed to be a remnant of a large moon that broke into pieces, creating it and Epimetheus (not seen here), often referred to as twin moons. Even seen at this distance of 136,000 miles (218,000 km), its ancient cratered surface is obvious. Prometheus is another of Saturn's irregular moons. The small satellite (only 63 miles [102 km] across) is much less cratered than Janus, and is found at the inner edge of the F ring.

Flying saucers or Saturn's moons? This image shows the distinctive flying saucer shape of two of Saturn's small, irregular moons. These moons possess, much like Iapetus, a very defined equatorial ridge separating the moon's two hemispheres. Atlas (left) orbits at the outer ridge of Saturn's A ring and is a mere 20 miles (32 km) across. Pan (right) is the innermost of Saturn's moons and is roughly 12 miles (20 km) across. Pan can be found in the Encke gap, in the A ring.

Saturn: Deeper Still

Although the Cassini-Huygens mission — a mission that sought to deepen our understanding of a complicated and delicately beautiful system billions of kilometers from our home — has helped broaden our understanding of Saturn and its satellites, it has also whetted the appetite of researchers the world over. From witnessing the longest-lasting storm in our solar system, to flying through Enceladus' geyser, to piercing the smog of Titan, the insight that scientists have gained about this jewel of our planetary neighborhood has also left them with new questions — new puzzles to solve. What causes the geysers? What system creates the enigmatic Saturn hexagon? How do the small moons influence the rings of Saturn?

It is clear that, in order to provide humanity with a deeper understanding of our own creation and that of Earth, we need to give planets such as Saturn much more attention. But there are other planets in our far-reaching system that deserve more study as well. The New Horizons spacecraft is on its way to peer even more deeply into the outreaches of space. Although Pluto, long considered the ninth planet in our solar system, is, as of 2006, no longer considered a planet by the International Astronomical Union, it will be targeted for study by New Horizons.

Dedicated to my daughter Sara, whom I encourage to always "look up"; to my family; to Cliff Woodrow, for buying me my first telescope; and to Rob for being the voice of encouragement.

ACKNOWLEDGMENTS

The author would like to thank the contributions of the following sources that were consulted for this book: the Jet Propulsion Laboratory's Photo Journal Website, Carolyn Porco and her team at the Cassini Imaging Central Laboratory for Operations (CICLOPS) for their incredible hard work and fantastic images, *Astronomy* magazine, *National Geographic* magazine and the *Firefly Atlas of the Universe*. A special thank you also to Terence Dickinson for helping to develop my avid interest in astronomy, and to Lionel Koffler.

IMAGE CREDITS

Pages 4, 7, 8, 9, 10, 11, 12, 13, 14, 15, 22, 23, 24, 25, 45, 55, 73 (bottom)
courtesy NASA/JPL

Pages 16, 18, 19
courtesy NASA, ESA, J. Clarke (Boston University), and Z. Levay (STScl)

Pages 17, 20
courtesy Erich Karkoschka (University of Arizona) and NASA

Page 21
courtesy NASA and The Hubble Heritage Team (STScl/AURA)
Acknowledgment: R.G. French (Wellesley College), J. Cuzzi (NASA/Ames),
L. Dones (SwRI), and J. Lissauer (NASA/Ames)

Pages 26, 27, 28–29, 31, 32–33, 34, 35, 36, 37, 38–39, 40 (left), 42, 43, 44,
46, 52, 53 (left and center), 54, 55, 56, 59, 60, 60–61, 62, 62–63, 63, 64–65,
66–67, 68, 69, 71, 74, 75, 76, 77, 78–79, 80–81, 82, 83, 84, 85, 86, 87, 88,
89, 90–91, 91
courtesy NASA/JPL/Space Science Institute

Pages 30, 40–41, 49, 50, 51, 53, 58–59, (right)
courtesy ESA/NASA/JPL/University of Arizona

Page 47
courtesy NASA/JPL/USGS

Page 56 (bottom right)
courtesy NASA/JPL/GSFC

Page 70
courtesy NASA/JPL/University of Colorado

GLOSSARY

Ammonia: A colorless and odorless gas composed of nitrogen and hydrogen; NH_3.
ASI: Agenzia Speciale Italiana (the Italian Space Agency); the ASI is part of the joint Cassini–Huygen's mission to Saturn.
Aurora: Glowing lights that appear at the planetary poles, caused by charged particles from the sun; it is believed that, because the particles are electrically charged, they are drawn to the magnetic poles.
Big Bang Theory: A theory that the observable universe was created at a single moment in time 14.7 billion years ago from an extremely hot and dense state in one single explosion.
Christiaan Huygens: (1629–1695) Dutch astronomer who discovered Saturn's moon Titan in 1655.
Convective upwelling: The process in which gas moves up through a planet's atmosphere from a hot region into a cooler region.
Earthshine: The faint lighting of the night side of the moon as a result of sunlight reflected from Earth to the moon.
Edwin Hubble: An American astronomer who discovered that galaxies are "island universes," not nebulae; he also developed a classification system for galaxies and the "Hubble Constant," which is the means that astronomers use to determine the distance of a galaxy.
Ejecta: Material that has been thrown out of a crater during an impact event.
Equatorial ridge: A raised area of a planet or moon that runs along its equator; in the Saturnian system, this feature is prominent on Iapetus, Atlas and Pan.
Equinox: The time when the sun crosses the celestial equator from north to south or south to north; this occurs twice a year on Earth, and is known as the vernal equinox (spring) and the autumnal equinox (autumn).
European Space Agency: A space agency consisting of several European countries; a major contributor to the Cassini–Huygens mission.
Galileo Galilei: (1564–1642) Italian physicist and astronomer who made improvements to the telescope and discovered four of Jupiter's moons (called the Galilean satellites); he is often referred to as the father of modern astronomy.
Great Red Spot (Jupiter): A circulating storm measuring more than the width of three Earths, located in Jupiter's upper atmosphere; this storm has lasted for at least 300 years.
Hubble Space Telescope (HST): A telescope that was launched in 1990 and orbits Earth.
Hydrocarbon: An organic compound that consists of hydrogen and carbon.
Infrared light: Electromagnetic radiation; infrared light has wavelengths that are longer than visible light but shorter than microwaves.
Mare: A large circular plain; literally, "sea."
Methane: An odorless gas that is produced by decomposing organic material.
Micrometeorites: Tiny meteorites.
Moonlet: A very small body orbiting a planet.
Near Infrared Camera and Multi-Object Spectrometer (NICMOS): A scientific instrument aboard the Cassini spacecraft that observes in infrared light.
Solstice: The time where the sun reaches its highest (summer) or lowest (winter) point in the sky at midday.
Spectroscopy: In astronomy, the method used to measure the light traveling from an object to Earth as well as the object's chemical composition; spectroscopy breaks visible light into different wavelengths.
Sulci: (Latin) furrows; ridges.
Tectonic activity: Activity on a body that is a result of the movement of its crust, or tectonic plates.
Transit: The passage of one celestial body in front of another.
Ultraviolet imaging spectrograph: An instrument aboard the Cassini spacecraft that is used to measure ultraviolet light — electromagnetic radiation that has wavelengths shorter than those of visible light.

INDEX